ok 6)

ALSO BY THE SAME AUTHOR:

AUDIO

(Elements of Electronics – Book 6)

by

F. A. WILSON

C.G.I.A., C.Eng., F.I.E.E., F.I.E.R.E., F.B.I.M.

BERNARD BABANI (publishing) LTD
THE GRAMPIANS
SHEPHERDS BUSH ROAD
LONDON W6 7NF
ENGLAND

PLEASE NOTE

© 1985 BERNARD BABANI (publishing) LTD

First Published — February 1985
Reprinted — July 1990

British Library Cataloguing in Publication Data
Wilson, F.A.
    Elements of electronics — (BP111)
    Book 6: Audio
    1. Electronic apparatus and appliances
    I. Title
    621.381   TK7870

    ISBN 0 85934 086 4

Printed and Bound in Great Britain by Cox & Wyman Ltd, Reading

PREFACE

There is in souls a sympathy with sounds;
And, as the mind is pitch'd the ear is pleas'd.
                                        William Cowper

Book 5 considers the elements of communication in a broad fashion. Because the subject is so wide-ranging it is only practicable there to look at the most important principles for a general technical appreciation of the subject as a whole. This present book examines a particular aspect of the communication scene to a greater depth so that the reader can develop a far more detailed and intimate appreciation of what is involved. A glance at the "Contents" shows the wide range covered.

For readers who are following the whole series as a course of study, the book is designed to be read from cover to cover. For those moderately or even highly skilled in the art, it is invaluable for updating or as a book of reference, picking out for explanation as it does those elements which all too often get mislaid in the memory. Generally in this series we work things out for ourselves from first principles without getting too involved with the more complicated mathematical formulae which tell the expert much but us very little. Nevertheless we recognize that the person who can fully understand the intricacies of electronics without recourse to some arithmetic and mathematics has yet to be born. Thus a certain amount of basic technical skill is assumed such as gained from the earlier books in the series (especially Books 1, 2 and 3) or as possessed by intermediate level students or electronics hobbyists.

For those readers who have some or all of the earlier books there is assistance in refreshing the memory by references which are made back to some of the more elusive technical points. The reference is raised from the text and is in the form Book/Section, for example $(2/4.1)$ means that additional information is contained in Section 4.1 of Book 2.

Brief descriptions of the earlier books in the series follows:

BOOK 1 "The Simple Electronic Circuit and Components". This deals with electricity, the electric circuit, electrostatics and electromagnetism backed up by several appendices teaching the required mathematics from arithmetic (for absolute beginners) through decimals to logarithms, simple mathematical equations and geometry.

BOOK 2 "Alternating Current Theory". Perhaps less mentally exhilarating but a must for all who wish to really get to grips with the subject. It considers the sine and more complex waveforms and how they react in the many basic alternating current circuits, also time constants together with the required mathematics, trigonometry and geometry.

BOOK 3 "Semiconductor Technology" prepares the way into the modern world of electronics with explanations of the working of semiconductors and of their practical characteristics, then of rectifiers, amplifiers, oscillators and switching (including a little on computers) with a chapter also on microminiature technology. The appendices cover the additional mathematics and binary arithmetic.

BOOK 4 "Microprocessing Systems and Circuits" branches out into the computer world. It is a book which really starts at the beginning of the whole subject and in this instance does not rely wholly on the preceding books, for non-electronically minded people there is a special appendix instead. A complete microcomputing system is explained bit by bit and both the need for and the execution of programming developed. This is followed by a comprehensive survey and explanation of the many basic electronic circuits in use. The appendices are written to provide an intimacy with numbering systems, especially binary.

BOOK 5 "Communication" assumes a modicum of basic electronics knowledge and mathematics. It begins with a discussion of some of the fundamental transmission aspects of communication including channel capacity and information flow. While not getting involved in the more complicated theory and mathematics, most of the modern transmission system techniques are examined including line, microwave, submarine, satellite and digital multiplex systems, radio and telegraphy. To assist in understanding these more thoroughly, chapters on signal processing, the electromagnetic wave, networks and transmission assessment are included. Finally, a short chapter on optical transmission.

CONTENTS

CHAPTER 1. THE SOUND WAVE

Audio is the Latin word for "I hear" and thus is defined the scope of this book, things electronic affecting what we hear. However, for completeness and full understanding many "non-electronic" features are also brought into the discussion for audio communication is seldom achieved without them, the simple air-path and the way rooms treat sound waves are just two examples.

In this first chapter we get to know the *sound wave* remembering that in fact it does not become sound until it has entered our ears and produced a sensation within.

1.1 VIBRATION

Vibration needs no introduction, it is simply any rapid to and fro or reciprocating motion. Sound waves arise almost invariably from mechanical vibration, the term "vibration" is derived from a word meaning to shake or swing and it is to the swing of a clock pendulum that we turn for a reminder of *simple harmonic motion.* (2/1.1) The time of swing of a pendulum depends on its length and in Fig.1.1 let the pendulum have such a length as to give a time for one complete oscillation of one second.

The graph below the pendulum is that which would be traced out by an inked brush on the bob in contact with a sheet of paper moving at constant speed. It is the familiar *sine* wave and the pendulum swing is in *simple* harmonic motion because its graph has the least complex waveform of all.

The *amplitude* and the *period* (T) of the wave are both shown in the Figure and in this case T = 1 sec., therefore the frequency of the wave, that is, the number of times a complete to and fro motion occurs in 1 sec. is

$$f = \frac{1}{T} \text{ Hz, in this case 1 Hz.}$$

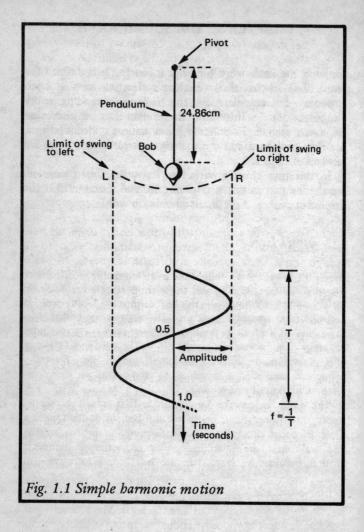

Fig. 1.1 *Simple harmonic motion*

A pendulum swinging is not heard because air flows round it as it moves. It is when the frequency of oscillation is sufficiently high (say above 20 Hz) that the air has no time to do this and its particles become alternately compressed and rarified. Let us move upwards in vibration frequency to that of, say, a tuning fork (f for an "A" tuning fork = 440 Hz). When struck the vibration may not be evident visually but it can be felt if either of the prongs is touched. The fork is an example of a mechanical vibrating system and as such it has a natural resonance causing it to vibrate best at one particular frequency. We shall encounter the phenomenon of mechanical resonance frequently so it is important to understand how it arises and fortunately we have a parallel with electrical resonance[2/3.7] as a guide.

Mechanically moving parts have *compliance, stiffness* and *mass*. We may be meeting some of these parameters for the first time and if so acquaintance with mass and *elasticity* ought first to be gained from reading Appendix 1 (A1.1 and A1.2). Springs in common with other elastic materials have compliance, that is, the degree to which they yield to an applied force. The opposite is stiffness, the degree to which they do *not* yield. We could take as an example the difference between mattress springs and those supporting a car body on its wheels. In comparison the car springs are less compliant because they have to carry a greater weight, they therefore possess greater stiffness, the particular characteristics of both types of spring being chosen to suit the weight to be carried. The tuning fork itself also provides an example for thick prongs have much greater stiffness than thin ones.

A helical spring supporting a weight is shown in Fig.1.2. At rest the extension of the spring depends on the size of the weight. If the latter is lifted then allowed to fall it will bob up and down and continue to do so for some time, the whole system is in resonance but at a very low frequency. With a floppy spring (low stiffness) the weight moves slowly but over a large distance, with a tight or stiff spring the movement is more rapid but its excursions are less, so changing from high amplitude/low frequency to low amplitude/high frequency. Frequency is therefore proportional to spring stiffness. Little imagination is required to see that if the weight or mass is increased the amplitude increases and the frequency falls. The electrical analogues of stiffness and mass are shown in the Figure and it is compliance which is equivalent to capacitance,

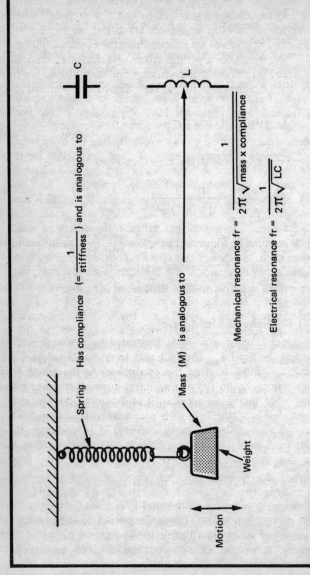

Spring   Has compliance $\left(= \dfrac{1}{\text{stiffness}}\right)$ and is analogous to

Mass (M) is analogous to

Mechanical resonance $fr = \dfrac{1}{2\pi \sqrt{\text{mass} \times \text{compliance}}}$

Electrical resonance $fr = \dfrac{1}{2\pi \sqrt{LC}}$

Fig. 1.2 Simple mechanical vibrating system

4

not stiffness. Compliance is a measure of the distance of movement caused by a given force and it is expressed in units of metres per newton (m/N). In acoustical engineering some relatively small values are encountered, for example, 0.01 m/N (10^{-5} cm/dyne) for a pickup stylus. Mass has its equivalence in inductance. Taking the analogy one step further we find that the requirement for a high resonance frequency is:

$$\text{electrically a low L x C product} \left(fr = \frac{1}{2\pi\sqrt{LC}} \right)$$

∴ mechanically a low mass x compliance product

$$\left(fr = \frac{1}{2\pi\sqrt{\text{mass x compliance}}} \right)$$

which can also be expressed as a low $\dfrac{mass}{stiffness}$ ratio or a high $\dfrac{stiffness}{mass}$ ratio. These ratios are important in the design of microphone and loudspeaker diaphragms (Chapter 5).

1.1.1 Generating a Sound Wave

The generation of a wave is illustrated by the sequence of drawings in Fig.1.3. The fact that in (i) the small arcs are equidistant indicates that the pressure along the length l of air is constant. In (ii) one vibrating tuning fork prong has compressed the air adjacent to it as shown by the arcs being drawn closer together. The air is compressed because, having inertia (i.e. inability to move quickly) it is unable to flow around the tuning fork prong in time. As an example of the time involved, with a 440 Hz tuning fork the prong moves from its central rest position to that shown in $\dfrac{1}{4}$ x $\dfrac{1}{440}$ secs., i.e. in just over half a millisecond. All gases (and air is one) resist a change in volume hence the pocket of compressed air expands and compresses the air to the right of it, this in turn does the same and so the compression travels to the right.

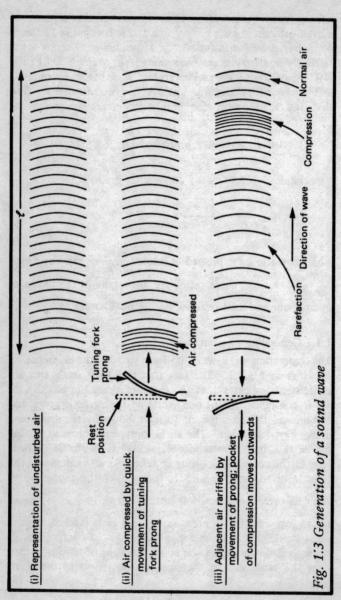

(i) Representation of undisturbed air

ℓ

Rest position

Tuning fork prong

(ii) Air compressed by quick movement of tuning fork prong

Air compressed

(iii) Adjacent air rarified by movement of prong: pocket of compression moves outwards

Rarefaction

Direction of wave

Compression

Normal air

Fig. 1:3 Generation of a sound wave

6

Because of its elasticity the prong now moves over to the left, eventually reducing the pressure on the air to such an extent that it gives rise to a pocket of rarefaction as shown in (iii).

Continuation of this process causes a series of alternately compressed and rarefied waves travelling outwards from the tuning fork. Particles of air in the path of the wave are caused to shift rapidly forwards and backwards so the original vibration is impressed on them. Because the particles move to and fro in the direction in which the wave is travelling, the type of wave is known as *longitudinal* (running lengthwise). Two points are especially worthy of note: (i) the air particles do not travel along with the wave, for still air they only vibrate about their rest (undisturbed) positions; (ii) for the transmission of sound waves some *medium* (gas or solid) is needed, they cannot travel in a vacuum.

The tuning fork used as an example in the preceding section has a *natural* frequency of oscillation. The metal prongs have elasticity, if one is displaced from its rest position and then released it returns to normal and because of its momentum, overshoots. The greater the overshoot, the greater the restoring tension in the metal hence the overshoot is limited and the prong comes to rest. It immediately springs back towards the normal position and with its momentum onwards in the direction of the original displacement. Continuation of this process is vibration but inevitably without help the vibration dies away. Looking at this a little more closely it is evident that the tuning fork (or anything else vibrating) has energy imparted to it by the displacing force and we can look upon such a vibrating system as having a certain $Q^{(2/3.4)}$ where $\dfrac{1}{Q}$ represents the fraction of energy lost per cycle through friction both within the material and due to the work done in moving the surrounding air particles. The similarity with the electrical oscillating circuit begins to show for in the latter Q also is related to energy losses. If a certain fraction of the energy is lost from the vibrating tuning fork per cycle then the actual amount lost is proportional to that available at the beginning of the cycle. This is sufficient reason to suspect that an exponential function[1/A4.1.2] is involved. A graph indicating what might happen to a tuning fork is shown in Fig.1.4. It is known as a *damped vibration or oscillation* and in this particular example the amplitude of the motion of the prongs falls from its maximum level to about one-tenth in just under 5 seconds. Clearly in such a drawing the vibration at 440 Hz cannot be portrayed, we manage with a few cycles only. The vibration amplitude follows an equation having two components, cos ωt which is the normal wave at frequency $f = \frac{\omega}{2\pi}$ multiplied by the decay factor $e^{-\gamma t}$ (γ is the Greek letter gamma). γ is a complex parameter but the point of importance to us is that it is inversely proportional to Q, thus as Q increases γ falls and $e^{-\gamma}$ increases[2/A4.2] meaning simply that high Q results in a lower rate of decay (damping) and vice versa, a not unexpected result especially in view of our experience with Q of the electrical circuit.

1.1.2 Resonance

Resonance occurs as in electronic circuits when sufficient energy is imparted at the natural frequency to the vibrating material to make up for the losses. The energy should preferably but not necessarily be supplied on each cycle and as we have found with the electrical counterpart, it must be in phase[2/3.7] for if otherwise it damps rather than assists the vibration. The organ pipe provides an example and we will see in Section 3.2.1 that an enclosed volume of air in a pipe has its own particular period of vibration (i.e. has its own natural frequency) so giving the particular note for the pipe. Accordingly when pressure changes are applied to the air in the pipe at its natural frequency, the air vibration is reinforced, resonance exists and the note sounds loudly (resonance is derived from the Latin word resonare, to resound).

Studies were made of this phenomenon way back in the early eighteen-hundreds when Hermon von Helmholtz, a German physicist and mathematician made a range of "sonorous globes" with small openings. Not only did each of these respond to its own particular frequency but he also

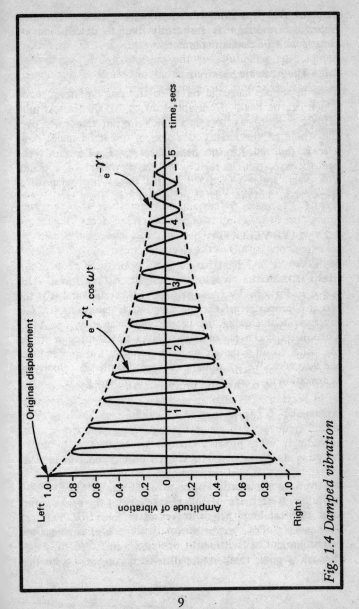

Fig. 1.4 Damped vibration

detected that harmonic vibrations were present. The term *Helmholtz resonator* is frequently used to describe an enclosure with air resonating in it.

1.1.3 The Acoustic Spectrum

This, in simple language refers to the range of frequencies which can be heard, from about 20 to 20,000 Hz. As individuals some of us have ears which are not responsive over the whole of the range, discussion of our deficiencies however is reserved for the next chapter. Fig.1.5 shows the acoustic spectrum in relation to the keyboard of a piano. The piano scale is used for getting things into perspective, it is explained in more detail in Section 8.1.2.

1.2 WAVE VELOCITY

Radio waves and light have a fixed *velocity of propagation* irrespective of the medium through which they travel. This is not so for sound waves, their speed is variable and it is the physical properties of the medium which determine it. The more *resilient* (having the property of elasticity, see A1.2) a medium is, the faster sound waves travel through it; on the other hand, a denser or more compact medium reduces the velocity. We are mostly concerned with the velocity of propagation in air and the basic formula for any gas is

$$\text{Velocity of propagation, } c = \sqrt{\frac{\gamma p}{\rho}}$$

where γ is the ratio of the specific heat at constant pressure to that at constant volume,

$\quad\quad$ p is the steady pressure of the gas in N/m^2 (1/A9.2)

$\quad\quad$ ρ (Greek letter rho) is the density of the gas in kg/m^3

(see A1.1 for extra help with regard to force and pressure).

$\quad$ γ needs a little explanation. *Specific heat* is a measure of the amount of heat required to raise a mass of a substance through a given temperature difference compared with that

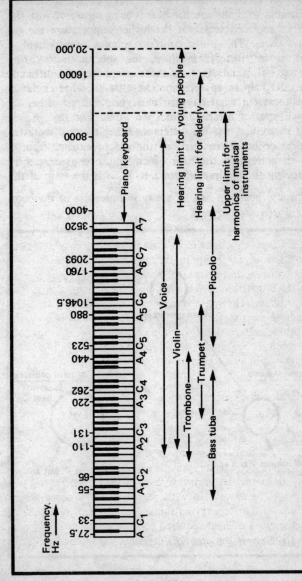

Fig. 1.5 Acoustic spectrum

11

required for an equal mass of water. Gases expand noticeably when heated so if the specific heat is being measured with the gas in an enclosed vessel, at the higher temperature the gas pressure rises. This is the *constant volume* measurement, a second is at *constant pressure*, the specific heats being designated C_v and C_p, so there are in fact two different values. To help in appreciating the difference the circles in Fig.1.6 represent what happens to a given volume of gas for the two types of measurement. We would find that C_p is greater than C_v and the difference is the external work (in the form of heat) required for the gas to expand from V to V′ or conversely the heat which would be generated by compressing the volume V′ down to V. γ is the ratio of the two specific heats $\dfrac{C_p}{C_v}$ and in a way is a measure of the com-

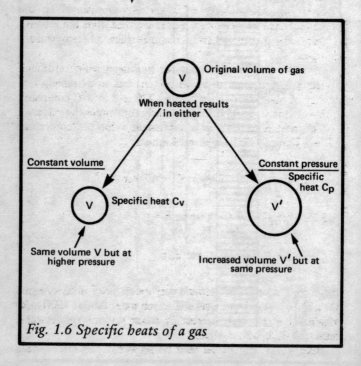

Fig. 1.6 Specific heats of a gas

pressibility of the gas. From the formula, the higher the value of γ, the greater is the sound wave velocity. For air γ is 1.403 and taking the "normal" steady pressure at the Earth's surface, p as 101333 N/m^2 (equal to 760 mm of mercury) and the density of air at $0°C$, ρ as 1.2927 kg/m^3,

$$c = \sqrt{\frac{1.403 \times 101333}{1.2927}} = 331 \text{ m/s}$$

This is a reference value at $0°C$, for other temperatures various formulae have been derived and experimentally verified, e.g., $c = 331.45 + 0.607T$ m/s where T is the temperature in $°C$ above 0 (up to $20°C$) from which at $20°C$, $c = 344$ m/s (approx. 1130 feet/sec.), an average figure for general use.

Note that the density of a gas is proportional to the pressure, hence as p changes, ρ does likewise, thus the velocity is less sensitive to pressure changes than to temperature changes.

Of interest is the fact that for hydrogen which although having γ of the same order as for air, has an extremely low density of about 0.09 kg/m^3, $c = 1270$ m/s at $0°C$ compared with that for air at 331, that is, nearly four times the velocity.

For most liquids there is a constant value of "compressibility", K, usable in the general formula

$$c = \sqrt{\frac{1}{K\rho}} \text{ m/s where } K = 47.9 \times 10^{-8}$$

and for example, for fresh water $\rho = 1$

$$\therefore c = \sqrt{\frac{10^8}{47.9}} \doteq 1,445 \text{ m/s}$$

therefore a sound wave travels very much faster in fresh water than in air and in fact faster still in sea water (about 1500 m/s) which is of especial interest in *sonar* systems (underwater detection using sound waves).

With solids, the shape may be important because dimensions could be commensurate with the wavelength λ, $\left(\lambda = \frac{c}{f} \right.$ as in radio,[2/1.2.1]$\left. \right)$ which complicates matters considerably. In bar-shaped steel the wave velocity is of the order of 5,000 m/s, a high velocity indeed compared with that in air. In lead the velocity is around 1250 m/s showing clearly the effect of its greater density.

1.2.1 Doppler Effect

Something which for many of us is an everyday experience is the *Doppler Effect* (after the Austrian mathematician and physicist, Christian Doppler). It relates to the apparent change in frequency of a sound when the source and the listener have movement relative to each other. [We will see in the next chapter that here we ought really to be talking in terms of the *pitch* of the sound, but this is of minor importance in this particular aspect.] The Doppler effect is most evident when sounds are being generated from fast moving vehicles, for example, the whistle of an express train, sirens on a police car or ambulance in which cases the source is moving but the listener is stationary. When the source is approaching, the peak of each sound wave arrives a little earlier than is usual and the apparent frequency increases, when the source recedes, the apparent frequency falls. Suppose the source is emitting a sound wave at f Hz. If stationary, there would be f waves in the distance travelled by sound in one second, that is, c metres. However, if the source is moving at v m/s, it travels v metres towards the listener in one second, hence the f waves are compressed into a distance $c - v$ metres

$\therefore$ Apparent wavelength, $\lambda_a = \dfrac{c - v}{f}$ m   from which,

Apparent frequency, $f_a = \lambda_a = \dfrac{c}{c - v} \cdot f$ Hz.

Similarly if the source is receding from the listener;

$$f_a = \frac{c}{c + v} \cdot f \text{ Hz.}$$

To appreciate this in a practical way, suppose a police car is travelling directly towards us at 70 km/h (just under 45 mph) and sounding its siren,

14

$$\text{then, } 70 \text{ km/h} = \frac{70 \times 10^3}{3600} = 19.44 \text{ m/s} (= v)$$

and fractional rise in frequency $= \dfrac{c}{c-v} = \dfrac{344}{344 - 19.44} = 1.0598$

and we will see later (Section 8.1.2 and Fig.8.1) that this is almost exactly a semitone rise (e.g. on the musical scale B to C, E to F or to the sharp of any note). Note however that when the car travels away from us at the same speed, the fall in frequency is less.

1.3 ACOUSTICAL QUANTITIES

In electronic engineering, when current flows it is normally within a wire, easy to contain and handle and it can be made to flow just where we wish. With the sound wave there is a world of difference for although built around the same framework of frequency and amplitude, the wave is free and if allowed, is ever expanding in all directions. Measurement of sound wave parameters is therefore more complex but not too difficult if we keep in mind some electrical equivalents with which we are more familiar. Two parameters of the sound wave which have parallels in electronics and are of considerable use in assessment of the wave are the *sound pressure*, p and the *sound intensity*, I.

1.3.1 Sound Pressure

It is obvious that compressions and rarefactions of a passing sound wave are able to exert an alternating force on a membrane or diaphragm. Now the force is not exerted at a point but over an area and sound pressure is therefore expressed as force per area in a plane usually at right angles to the direction of travel of the wave. Nowadays the unit is Newtons per metre^2 (N/m^2). Being an alternating quantity we must state whether the measurement is peak, r.m.s., or some other description of the waveform, much depending on whether it has constant amplitude or as with speech or music is continuously varying in form, amplitude and frequency. We will not complicate this section with the latter features but merely look at things generally.

Sound pressure is a measure of the fluctuation which the wave causes above and below atmospheric pressure. Its parallel in electronics is electrical pressure, that is, voltage. It is more convenient to use the logarithmic unit, the decibel[5/2.1] when quoting sound pressures (the ear has logarithmic tendencies), thus a generally accepted reference level, p_o is required. We use the sound pressure which is at approximately the *threshold of audibility* for sound at 1000 Hz, that is, the pressure exerted by a 1000 Hz sound wave on the ear drum which is just sufficient for the sound to be heard (hearing is considered in more detail in the next chapter). For audio acoustics the pressure is taken as 2×10^{-5} N/m^2, this value has become an internationally agreed standard although a slightly different value of 2.04×10^{-5} N/m^2 may be found in some publications. To get this reference level into perspective we would find that a whisper from one or two metres away would create a sound pressure at the ear of a listener at least ten times greater.

Readers who have already studied this subject may recall another sound pressure unit, the dyne per square centimetre, using this reference pressure, p_o is 2×10^{-4} dyn/cm^2, i.e. 0.0002 dyn/cm^2 (also see Appendix A1.1).

The *sound pressure level* is therefore defined as

$$20 \log_{10} \left(\frac{p}{p_o} \right) dB$$

where p is the actual and p_o the reference pressure in N/m^2. Note that the formula uses 20 log instead of 10 log because pressure is akin to voltage, not power. Because it is in such frequent use in acoustics, the abbreviation s.p.l. has become standard.

We look at typical practical sound pressure levels in Section 1.3.3.

1.3.2 Sound Intensity

The *intensity* of a sound wave indicates the acoustical power flow over a given area and as with pressure the area is considered to be in a plane at right angles to the direction of wave travel. As in the electrical circuit in which power varies as the square of the voltage so acoustic power varies as the square of the pressure and the relationship between them is given by

Sound Intensity, $I = \dfrac{p^2}{\rho c}$ watts per square metre (W/m^2) where

p is the sound pressure in N/m^2
ρ is the *density* of the medium in kilogrammes per cubic metre (kg/m^3)
c is the velocity of propagation in the particular medium in metres per second (m/s)

The sound intensity at the threshold of audibility, I_o is 10^{-12} W/m^2, hence

$$\text{Sound intensity } Level = 10 \log_{10} \left(\frac{I}{I_o} \right) dB$$

where I is the actual and I_o the reference intensity in W/m^2.

Because p_o and I_o both refer to the same reference (the threshold of audibility), measurements on any particular sound wave give numerically the same result for both pressure and intensity levels, that is, a sound pressure level of x dB is equivalent to a sound intensity level of x dB. An example may help to clarify this. Suppose the sound intensity generated by a motor car is 10 $\mu W/m^2$, i.e. 10^{-5} W/m^2. In decibels therefore above the reference level of 10^{-12} W/m^2 (= 0 dB)

$$\text{Sound intensity level} = 10 \log \frac{10^{-5}}{10^{-12}} = 10 \log 10^7 = 70 \, dB$$

(nearly always positive because levels below 0 dB are not heard).

From this can be found the sound pressure, p because the s.p.l. also equals 70 dB.

$$\text{Therefore since s.p.l.} = 20 \log \frac{p}{p_o} \, dB$$

$$\therefore 70 = 20 \log \frac{p}{2 \times 10^{-5}} \quad \therefore \frac{70}{20} = \log \frac{p}{2 \times 10^{-5}}$$

$$\therefore \text{antilog } 3.5 = \frac{p}{2 \times 10^{-5}}$$

$$\therefore p = 3162 \times 2 \times 10^{-5} \text{ N/m}^2 = 0.06324 \text{ N/m}^2.$$

Alternatively we could use the formula linking intensity and pressure. Considering air as the medium:

$$I = \frac{p^2}{\rho c}$$

I in this case being 10^{-5} W/m^2. ρ for air ≈ 1.2 kg/m^3. $c \approx 344$ m/s.

Then $p = \sqrt{I \rho c}$ N/m$^2 = \sqrt{10^{-5} \times 1.2 \times 344} = 0.06425$ N/m^2 the answers not being in exact agreement because of the approximations made.

1.3.3 Practical Sound Levels

To bring the formulae in this Section (1.3) together and add a touch of realism to what might seem almost meaningless figures, Fig.1.7 makes a composite presentation of sound pressure, intensity and decibel levels together with practical examples. The latter show the order of things generally, none is precise, even the thresholds of hearing and feeling (discussed in Chapter 2) vary widely from person to person. Nevertheless it is useful to gain some experience with the figures especially those for sound pressure level because it is this which is used most frequently in measurements.

1.3.3.1 Effects of Distance.

For simplicity sound pressure and intensity have been discussed for the *plane* wave. Waves can be considered as plane at some distance from the source but in practice at short distances the wave is more *spherical*, in fact with a point source and no obstruction (known as *free-field* radiation) it

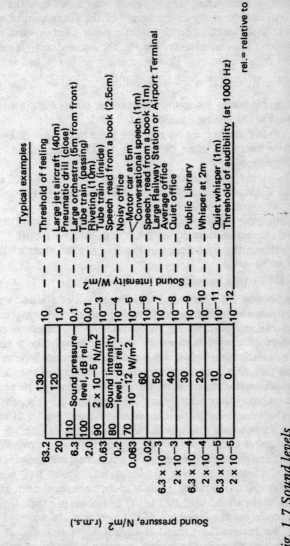

Fig. 1.7 Sound levels

Sound pressure, N/m² (r.m.s.)	Sound pressure level, dB rel. 2×10^{-5} N/m²	Sound intensity level, dB rel. 10^{-12} W/m²	Sound intensity W/m²	Typical examples
63.2	130			— Threshold of feeling
20	120		10	— Large jet aircraft (40m)
6.3	110		1.0	— Pneumatic drill (close)
2.0	100		0.1	— Large orchestra (5m from front)
0.63	90		0.01	— Tube train (passing)
0.2	80		10^{-3}	— Riveting (10m)
0.063	70		10^{-4}	— Tube train (inside)
0.02	60		10^{-5}	— Speech read from a book (2.5cm)
6.3 × 10⁻³	50		10^{-6}	— Noisy office
2 × 10⁻³	40		10^{-7}	— Motor car at 5m / Conversational speech (1m)
6.3 × 10⁻⁴	30		10^{-8}	— Speech, read from a book (1m)
2 × 10⁻⁴	20		10^{-9}	— Large Railway Station or Airport Terminal
6.3 × 10⁻⁵	10		10^{-10}	— Average office
2 × 10⁻⁵	0		10^{-11}	— Quiet office
			10^{-12}	— Public Library
				— Whisper at 2m
				— Quiet whisper (1m)
				— Threshold of audibility (at 1000 Hz)

rel. = relative to

is truly spherical. This can be appreciated from Fig.A1.1 in Appendix A1.3. The appendix also shows how the power (W) varies inversely as the square of the distance when energy is radiated in all directions from a point source, worth noting because it demonstrates how rapidly a sound wave is attenuated on its journey. However, since $W \propto p^2$ and therefore $p \propto \sqrt{W}$, then if $W \propto \dfrac{1}{d^2}$, it follows that $p \propto \dfrac{1}{d}$, that is, the sound pressure follows not the inverse square (of the distance) law but the inverse distance law.

Proof that we are on the right track can be gained by using these two laws when distance is, for example, halved. When this happens:

(i) the sound intensity, I, increases 4 times, in decibels the sound intensity level is therefore increased by 10 log 4 = 10 x 0.6020 = 6.02 dB;

(ii) the sound pressure, p, is doubled, the s.p.l. is therefore increased by 20 log 2 = 20 x 0.3010 = 6.02 dB.

Hence the equivalence of the levels in the chart of Fig.1.7 remains intact.

1.3.4 Particle Velocity and Displacement

In Section 1.1.1 we noted that the action of a sound wave in its passage through a gas is to impose an oscillating motion on the gas particles or molecules.[1/1.1] A measure of the effect of a wave is therefore the *particle velocity* it creates. The relationship between this and sound pressure follows from the fact that it is the movement or velocity of the particles which creates the pressure, for with no movement, no pressure is set up. Thus velocity and pressure are directly related as we will see from the formula which follows. Fig.1.8 illustrates the relationship for one cycle of a pure note and we must remember that we are considering particle, not wave velocity. At the top of Fig.1.8 are seven small diagrams showing steps in the oscillation of a single particle about its "normal" (undisturbed) position (indicated by a vertical dotted line). The steps are labelled (i) to (vii) and occur in a time sequence according to the frequency of the passing sound wave.

We assume the particle is already in motion and at (i) is travelling through the normal position. This is equivalent to

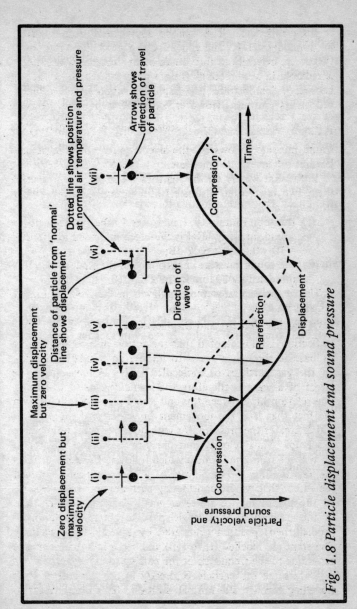

Fig. 1.8 Particle displacement and sound pressure

the lowest point of the swing of a pendulum where it moves through its rest position but in fact where the velocity is greatest. This also occurs at (v) and (vii) although at (v) the direction of travel is opposite.

Position (iii) is equivalent to a pendulum at its maximum swing where it must pause before changing direction. At the pause its velocity is zero. Thus as shown the particle has maximum displacement but zero velocity. Position (vi) is similar and positions (ii) and (iv) show the conditions at intermediate times. The full line graph below representing both velocity and sound pressure, when compared with the dotted one for displacement shows more clearly the 90° phase difference. The formula linking them is

$$\text{Particle velocity, } u = \frac{p}{\rho c} \text{ m/s}$$

where p = sound pressure in N/m^2
ρ = density of gas in kg/m^3
c = velicity of wave in m/s.

Using as an example an s.p.l. of 70 dB, equivalent to a sound pressure of 0.063 N/m^2 (Fig.1.7) in open air at 20°C, $p = 0.063$ N/m^2, $\rho = 1.2$ kg/m^3, $c = 344$ m/s. Then

$$u = \frac{0.063}{1.2 \times 344} = 1.526 \times 10^{-4} \text{ m/s}$$

in effect a small fraction of 1 mm per second.

The *particle displacement*, also known as *particle amplitude* is the product of particle velocity and time. Because the period of a wave varies inversely as its frequency, then displacement varies similarly. The formula is

$$\text{Displacement, } d = \frac{u}{2\pi f}$$

where u = particle velocity in m/s
f = frequency in Hz,
the transmission of lower frequencies therefore requires greater displacements.

For the same sound pressure as above at 1000 Hz pure tone

$$d = \frac{1.526 \times 10^{-4}}{2\pi \times 1000} = 2.4 \times 10^{-8}\,\text{m} \quad \text{or} \quad 2.4 \times 10^{-5}\,\text{mm.}$$

The sound pressure from which we calculated is r.m.s. we are often interested in the peak—peak displacement to show the maximum swing of the particles, which is therefore $2.4 \times 10^{-5} \times 2\sqrt{2}$ ($\sqrt{2}$ is the peak/r.m.s. ratio of a sine wave),[2/1.3.3] i.e. about 6.8×10^{-5} mm, less than one ten thousandth of 1 mm, a very small shift indeed. At the upper end of the scale however for the threshold of feeling (Fig.1.7) at 40 Hz, the peak—peak displacement is more than half of 1 mm.

From the electrical point of view the particle velocity has a similarity with current therefore the product of the sound pressure p and particle velocity u gives the sound power;

$$I = p \cdot u$$

which brings us back to the earlier formula for sound power or intensity, namely

$$I = \frac{p^2}{\rho c}$$

from which, since

$$u = \frac{p}{\rho c}, \text{ then } I = p \cdot u$$

Following Ohm's Law in electricity where $\text{watts} = \dfrac{\text{volts}^2}{\text{ohms}}$
ρc is sometimes called the *characteristic impedance*[5/6.1.2] of the particle medium, the units being *acoustical ohms*. ρc is sensitive to air temperature and pressure but a value for air under normal conditions which has general acceptance is 407 acoustical ohms. The slightly more approximate value of 400 is important for then the reference sound intensity of

10^{-12} W/m^2 and reference sound pressure of 2×10^{-5} N/m^2 move into exact agreement for

$$I = \frac{p^2}{\rho c} \quad \text{becomes}$$

$$\frac{\left(2 \times 10^{-5}\right)^2}{400} = \frac{4 \times 10^{-10}}{400} = 10^{-12} \text{ W/m}^2 \; .$$

1.3.5 Summary

Undoubtedly the elements of acoustical engineering presented in this Section (1.3) are of inestimable value in understanding the mechanics of sound generators and transducers, thus it is useful here to focus on some of the more salient points.

(i) the *velocity of propagation* of a sound wave varies according to the medium through which the wave travels. It is usually air and for this the velocity under "normal" conditions is 344 m/s (Sect.1.2).

(ii) *sound pressure* is analogous to voltage in the electrical circuit. It is quoted in terms of force per unit area, nowadays in newtons per square metre (N/m^2). For convenience a logarithmic scale of sound pressure levels is usually employed where the level is stated in decibels above the reference level of 2×10^{-5} N/m^2.

(iii) *sound intensity* is a measure of the acoustical power flow per area in watts per square metre (W/m^2). It varies directly as the square of the sound pressure and inversely as the density of the gas (or material) and the velocity of propagation. As with sound pressure, a logarithmic scale is convenient.

(iv) it is the *particle velocity* set up by a sound wave which creates sound pressure. The velocity varies inversely as ρc, the product of gas density and wave velocity. In the analogy with the electrical circuit, ρc has some equivalence to impedance and since sound pressure has something in common with voltage, particle velocity can be considered as being equivalent to current for as seen in Sect.1.3.4, the velocity, operating within the acoustical impedance of the medium gives rise to pressure (by Ohm's Law, I x Z = V).

(v) *particle displacement* or *particle amplitude* is proportional to the particle velocity and to the wavelength, it is therefore inversely proportional to the wave frequency.

Fig.1.8 puts pressure, velocity and displacement into a pictorial form. At this early stage we may not see the practical relevance of such parameters as say, particle velocity, but this will become clear later for example, there is a certain type of transducer known as a *velocity microphone.*

1.4 THE EFFECTS OF AN OBSTACLE

Normally free-field conditions do not exist and a sound wave eventually meets some barrier or object in its path, then arise absorption, reflection and diffraction. Because of the unquantifiable variables involved the subject cannot be treated here with any degree of precision or analysis, nevertheless by using a commonsense approach the general principles will become evident.

1.4.1 Absorption

It is without doubt fortunate for us that all sounds die away owing to absorption. If a sound wave meets a surface which is porous, air is set in motion within the tiny cavities and energy is dissipated in their walls. If the material is fibrous, not only do the cavities dissipate energy but the fibres may also be forced into vibration, again absorbing energy from the wave. In addition whole surfaces may be set in vibration, for example, a thin wooden panel. The acoustical energy is dissipated either as heat or through transfer into the mechanical energy needed to set things in motion. Data on some typical sound absorbent materials are given in Fig.1.9. The *absorption coefficient* expresses the value of the material as a sound absorber, a coefficient of 1.0 indicates that all of the sound wave is absorbed, conversely 0 means that none is absorbed. Note from the Figure how the absorption coefficient of some materials varies significantly with frequency and also from the few characteristics shown the very large differ-

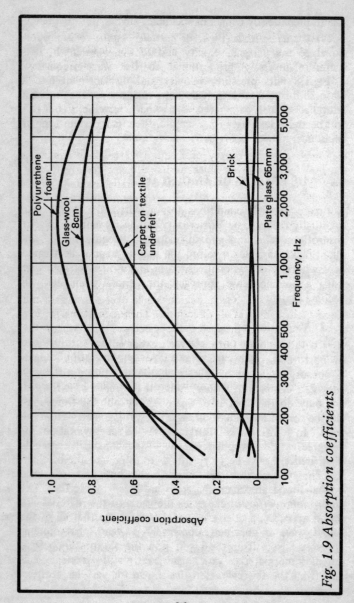

Fig. 1.9 Absorption coefficients

ences there can be. As an example, in the mid-frequency range polyurethane foam absorbs almost all of a sound wave reaching it, whereas plate glass absorbs practically none, this is to be expected for the smooth surface of plate glass is neither porous nor fibrous. From characteristics such as these the most efficient materials for the acoustic damping of rooms can be selected (see Chapter 4).

1.4.2 Reflection

It follows that the energy of a sound wave which is neither transmitted through an object in its path nor absorbed, must be reflected. There is a parallel with line transmission in which a signal travelling down an incorrectly terminated line is not completely absorbed and therefore experiences reflection[5/6.1.4]. Mathematical analysis of the electrical line is precise but if we try to do the same in the acoustic case we are in trouble because although standing waves also appear and there are nodes and antinodes of both pressure and displacement the sound wave which for example gives rise to an echo from a cliff reaches the obstacle over a large area and is reflected from it in all directions. These conditions are less clear-cut than are those for the electrical line where the reflected wave can only traverse exactly the path of the incident wave. However, a practical demonstration of the acoustical case can be given by projecting a pure-tone sound wave towards an object from which it is reflected back along the incident path. A listener is then able to experience the standing wave pattern along the path by the rise and fall in loudness as he or she moves between the source and the reflector. A microphone and measuring set instead would avoid subjective error.

1.4.3 Diffraction

In this section and the next we discuss the effects of objects within a sound field and it will be found that object dimensions relative to sound wave length are important. To see the outcome in a practical form it is worth recalling that at a velocity of 344 m/s

a frequency of	100 Hz	has a wavelength of 3.44 m
” ” ”	1,000 Hz	” ” ” ” 34 cm
” ” ”	10,000 Hz	” ” ” ” 3.4 cm

Diffraction can be considered in a rather simplified but nevertheless instructive manner as follows. Consider an obstacle having dimensions of the order of the wavelength of an incident sound wave. Some reflection occurs and under these particular conditions it takes place in all directions, the net result being to increase the pressure at the front of the object because both incident and reflected waves are present. Thus also arises an abnormal sound pressure distribution around the sides and at the rear. The object therefore interferes with the original direction of propagation of the wave by its effect on the pressure distribution. This is known as *diffraction* (from Latin, to break up) and it is of interest because the effect concerns microphones, loudspeakers and even ourselves when listening because there is diffraction around the human head.

1.4.4 Obstacle Effect

Once having passed an object the uneven pressure distribution tends to smooth out but naturally immediately to the rear of the object practically no wave at all exists. Using light terminology, we say that the object has cast a shadow, in this case an acoustic one. Now for a low frequency wave the time taken for a compression or rarefaction region to pass a given point is comparatively long (10 ms at 100 Hz), hence dispersion of the wave sideways into the shadow region is more easily accomplished than for a high-frequency wave in which compressions and rarefactions follow each other more rapidly. The result of this is that the length of the shadow behind the object is negligible for wavelengths considerably in excess of the dimensions of the object but significant when the wavelength is equal to or shorter than those dimensions. In the latter case the sound wave pressure behind the object is reduced considerably until eventually at some further distance behind it, dispersion from the sides at these frequencies "fills" the shadow. This is known as the *obstacle effect* and as a single example of its importance, objects should not be placed sufficiently closed to a microphone that the faithful reproduction of a sound is impaired by the microphone being within the acoustic shadow.

CHAPTER 2. HEARING

Having studied the sound wave in Chapter 1 we next see how
it is converted into the sensation we call sound. The mechanism is an entirely human function which is fascinating because
of its priceless sensitivity and performance, but which by
residing partly within the brain, leaves it for most of us,
shrouded in mystery. Nevertheless we can get to grips with
the elements of the system, leaving the study of what happens
within the brain itself to others.

2.1 HUMAN MECHANISM

The external section of the ear needs no description, it is a
rather peculiarly shaped arrangement and although mainly
decorative it does help in directing sound waves into the
auditory canal which is shown to the left in Fig.2.1(i). The
auditory canal disappears into the head and is the visual limit
for most of us but doctors can go further by using an
otoscope, an instrument for illuminating and examining the
eardrum. The Figure shows the essential features of the hearing mechanism from the auditory canal inwards. Within the
canal the pressure variations of the sound wave cause the
eardrum, a stretched membrane, to vibrate. Within the cavity
of the *middle ear* a chain of three tiny bones actually amplifies
the vibrations while transmitting them to the *inner ear*. To
non-medical people the bones are known as hammer, anvil
and stirrup as might be imagined from their shapes. From
the Figure it can be seen that the "handle" of the hammer is
attached to the centre of the eardrum. At the remote end of
the chain the stirrup transmits the vibrations to fluids within
the *cochlea* (Latin for "snail"), a unit through which the
vibrations are able to excite nerve endings of the *acoustic
nerve*. The *Eustachian tube* (after Eustachius, an Italian
anatomist) runs down from the middle ear for some 3—4 cms
into the throat so that the middle ear is open to atmospheric
pressure as is the auditory canal, thus the eardrum has equal

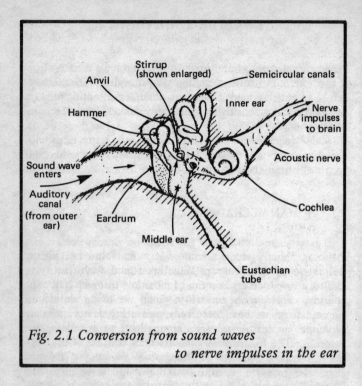

*Fig. 2.1 Conversion from sound waves
to nerve impulses in the ear*

pressure on both sides and can therefore vibrate freely. Generally the vibration amplitude of the eardrum is microscopic, down to as little as 10^{-8} mm is said to be detectable.

It is in the cochlea that things begin to get really complicated. The stirrup bone striking against it produces waves within the fluid. The cochlea is lined with a membrane containing thousands of hair cells tuned to vibrate at different frequencies, the oscillations in the fluid around the cells stimulate the appropriate ones to respond. The fibres of the acoustic nerve transmit these responses in terms of both frequency and intensity of the stimulation to the brain. The cochlea has about 2¾ turns and a total length of some 3 cm. The highest frequencies are effective at the stirrup end with the lowest at the centre of the spiral. The responses transmitted along the acoustic nerve become sound, precisely how the brain achieves this is still one of Nature's well-guarded secrets.

Semicircular canals are shown in the Figure, there are three of them lying in planes at right-angles to each other and they supply information to the brain about body position for balance, it is most unlikely that they take any part in the hearing process.

Our two ears function independently, there is practically no acoustic connexion between them. Tests have shown that from ear to ear through the skull there is an attenuation of some 50 dB.

2.2 AUDIBILITY

The next logical step is to find out more about hearing or *audibility* itself. In Chapter 1 the acoustical quantities considered are physical and so can be measured using instruments but when considering audibility which we might define as "perception by the ear", the only measuring devices are us, the human beings — and what a varied lot we are with our unpredictable requirements and different acuities of hearing. So, with human beings as the testing device we do not *measure*, we *assess*. This all boils down to the warning that for most data presented in this Section the result is simply the average for many people and that individual digressions from this average may be large. Nevertheless, from all the work carried out, we learn much. Tests using human beings as observers or listeners are described as *subjective* whereas for instrument measurement they are *objective*.

2.2.1 Pitch

Dictionaries give various definitions of *pitch* but perhaps the most meaningful for us is that it is the subjective quality of a sound which positions it in the musical scale. The immediate reaction is that this is in fact *frequency* but it is not, although pitch and frequency are closely related. Pitch is determined subjectively and if two pure tones of slightly different frequencies are presented to a listener and he or she is allowed to adjust the *intensity* of one of them, a level can be found at which the tones appear to have equal pitch. For a given frequency therefore, pitch is to a certain extent dependent on intensity. At low intensity levels pitch and frequency are virtually the same, at higher levels, say, above a sound intensity level of 40–50 dB (Fig.1.7), for low frequency tones the pitch decreases with increase in intensity but for high tones the pitch increases with increase in intensity. In the middle range of frequencies the effect is slight.

2.2.2 Loudness

Loudness is a term which relates to the magnitude of the sensation which a sound creates in the brain, it therefore depends mainly on sound intensity but because the ear is not equally sensitive at all frequencies it depends on frequency also. We must never forget that it is a *subjective impression*. We cannot ourselves assign an absolute value to the loudness of a sound so if confronted by two sounds at different levels to which we can listen separately, we are uncertain, except to say that one is louder than the other, by how much is left unanswered. But we can compare two different sounds and adjust the intensity of one of them until they appear to be equally loud. The technique is therefore to present listeners with the sound to be assessed and allow them to switch between this and a reference tone of 1000 Hz which is adjusted in intensity in order to find a level for "equal loudness". The sound pressure or sound intensity level of the tone then gives the *loudness level* of the sound wave in *phons*. Stated more precisely, the loudness level in phons is the sound pressure level or sound intensity level of a 1000 Hz reference tone which is judged to be equally loud.

The phon scale is one of decibels relative to the accepted threshold of hearing, it is logarithmic. An alternative scale of loudness uses the *sone* as the unit, it is a linear scale in that doubling or halving the sone value relates to doubling or halving the loudness. Thus we would consider a sound at a level of x sones to be x times louder than if its level were 1 sone. The internationally standardized relationship between loudness level expressed in phons and in sones is given in Table 2.1 and we see that doubling the sone value is equivalent to a rise of 10 dB in the phon value meaning that on average when human beings consider a sound to be *twice* as loud as another:

(i) that sound has an intensity 10 dB greater or 10 times as much as the other;

(ii) the sound pressure is also 10 dB greater but $\sqrt{10}$ times as much;

(iii) it has a phon value 10 higher or double the sone value.

(Help may be required perhaps from the scales of Fig.1.7.)

2.2.2.1 Minimum Perceptible Changes

So that decibels of loudness actually mean something to us when discussed later, a few suggestions relating them to how they are heard may be useful. These come from people with about average hearing. Using pure tones from a loudspeaker or earphone over the frequency range 50–10,000 Hz and a sound pressure level of 50 dB or more, a change of 1 dB can just be detected. At lower sound pressure levels, say, below 40 dB a change of 3 dB is needed before becoming perceptible, meaning that a pure tone changing by less than 3 dB would be considered by most people not to have changed at all. Fig.1.7 may help to keep the sound pressure levels in perspective.

2.2.3 Aural Sensitivity

If a listener is provided with a pure tone sound, usually through calibrated headphones and under very quiet ambient conditions and the level of the tone gradually reduced, there comes a point where at one level the tone can just be heard but when reduced only slightly, there is no trace of it. The tone has in fact dropped below the *threshold of audibility* for that person at that particular frequency. Even though instruments may indicate that the tone is still giving rise to a sound pressure, the listener will rightly insist that it cannot be heard. There is just not enough sound power to overcome the inertia of the ear vibratory system and get it moving. We can

LOUDNESS LEVEL	Phons	20	30	40	50	60	70	80	90	100	110	120
	Sones	0.25	0.5	1	2	4	8	16	32	64	128	256

Table 2.1 Relationship between loudness expressed in phons and in sones

therefore define the threshold of audibility for any particular sound as the minimum effective sound pressure capable of providing an auditory sensation. We each have our own personal thresholds but as we saw in Chapter 1, a standard or reference one has been measured, actually on young people with good hearing. The threshold varies with frequency, it is higher at frequencies below about 1000 Hz and above about 5000 Hz. It also rises with age, this is discussed later. Note that a *rise* in the threshold means that higher sound pressures are required for the sound to be heard at all. The standard threshold of audibility curve is given in Fig.2.2. Recalling from Sect.1.3.1 that the threshold of audibility at 1000 Hz is the generally accepted reference level, Fig.2.2 seems to disagree, the curve does not pass through 0 dB at 1000 Hz, it is some 4 dB higher. This discrepancy arises from the fact that the reference level was chosen from original work whereas later experiments have produced results which differ slightly. The curve shown is the present recommended one. Naturally the threshold rises for listening with one ear only, about 1.5 dB for frequencies below about 3000 Hz at frequencies above this up to almost 5 dB.

Also shown in the Figure are *equal loudness contours* for 40 and 80 phon (4 and 16 sone). These naturally pass through the 40 and 80 sound pressure levels at 1000 Hz by reason of the definition of the phon. Such contours for other loudness levels up to 120 phon have also been produced, the two shown are sufficient for us to get to grips with the principles and also happen to enclose an average sort of sound level range (see Fig.1.7). What is most striking about the threshold curve is that at low frequencies, taking, say, 50 Hz as an example, nearly 40 dB higher sound pressure is required for a sound to be heard than at 1000 Hz so we begin to appreciate that when listening to music played softly we could miss some of the lowest frequencies altogether. At a loudness level of 40 phon the difference is less (about 25 dB) and at 80 phon less still (about 15 dB). Hence the idea incorporated in many hi-fi systems of automatically raising the bass response of the system on low loudness control settings. A rise in the threshold also occurs at higher frequencies, particularly just below

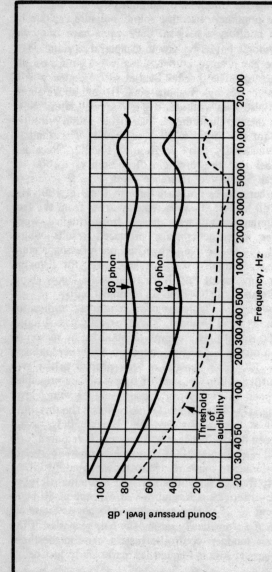

Fig. 2.2. Equal loudness contours

10,000 Hz, however, generally limited to about 10 dB.

There is also an upper limit, varying slightly with frequency at which the loudness is so great that people experience discomfort and even pain. Increases in sound pressure above this level are not detected because in a way, the ears are saturated. The limit is known as the *threshold of feeling* and it is more difficult to quantify because the levels at which subjects consider feeling or pain to occur vary widely from person to person. Several curves have been published but it is perhaps sufficient to us to remember that it occurs *around* 130 dB s.p.1.

2.2.4 Hearing and Age

Apart from those who unfortunately are born with poor hearing and those who lose some or all hearing through illness or accident, it is a rather disturbing fact that generally our hearing deteriorates with age. The loss is especially significant at the higher frequencies. When young our range is up to 20 kHz or higher, by the sixties most have difficulty in hearing sounds above 12 kHz. This high frequency loss is attributed to the gradual hardening of tissues in the hearing mechanism and their consequent loss of elasticity. Pure tone tests have established that as many as 25% of a population may have a threshold over 15 dB above the minimum (i.e. a *hearing loss* of 15 dB) and 10% over 25 dB above.

What matters to most people is clarity when listening to speech and this is the range used in the threshold curves of Fig.2.3(i) interpreted in terms of hearing loss in Fig.2.3(ii). Again the reminder that the curves are drawn from averages of small samples of the population so individual cases may deviate considerably.

The difference in hearing between the sexes is also of interest, it would appear from the few experiments so far carried out that on average teenage girls have slightly better hearing (1–2 dB) than boys and this superiority seems to be maintained throughout life for when aged between 50 and 60, although suffering an extra loss at low frequencies of a few decibels, there is a much smaller loss at the higher speech frequencies as indicated in Fig.2.3(ii).

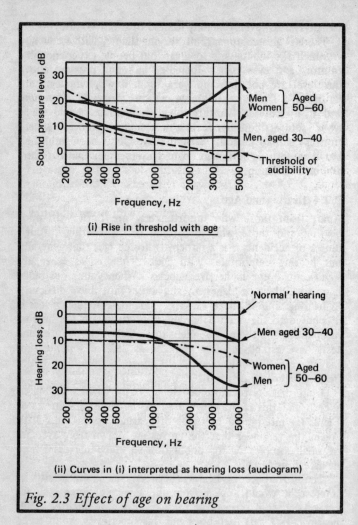

(i) Rise in threshold with age

(ii) Curves in (i) interpreted as hearing loss (audiogram)

Fig. 2.3 Effect of age on hearing

2.2.4.1 Audiometry

Most *audiometry* (hearing measurement) is conducted to estimate the minimum intensity of sounds which can be heard, over a range of frequencies. The results in graph form are known as *audiograms*. The technique is straightforward, using an *audiometer* to provide tones at different frequencies via an attenuator to a pair of earphones worn by the person whose hearing is being measured. Usually high quality moving coil earphones (see Chapter 5) with a wide frequency range are used. The whole system is calibrated against "normal hearing", for anybody with this the audiometer dial read 0 dB. Tones are switched to the headphones one ear at a time and reduced in intensity to the point at which the subject indicates that they can no longer be heard, the audiometer then shows the hearing loss at the particular frequency. The sound pressure required from the earphones for a very deaf person (say, with 80 dB hearing loss at 1000 Hz) is 80 dB higher than for a person with normal hearing, hence the earphones must be capable of handling large powers without distortion. An audiogram takes the form of Fig.2.3(ii) having two graphs, one for each ear.

2.3   HEARING AIDS

It was not until the very early nineteen-hundreds that the acoustic amplification of the ear trumpet gave way to electronics. Development in improving gain and frequency response and in reducing size has continued incessantly since then. The modern hearing aid is no different in basic principle from the early ones, simply a combination of microphone, amplifier, earphone and power supply for listening to the incoming sound and presenting it to the listener in amplified form. We study the principles of the transducers later, here we simply get acquainted with the device as an aid to poor hearing. Gain is no longer a problem but stability sometimes is so we look at the latter first.

2.3.1 Stability
In any amplified system where some of the output may find its way back to the input there is the possibility of instability. It is desirable for oscillators [3/3.3] but certainly not for hearing aids. In the elementary hearing aid arrangement of

41

Fig. 2.4(i) the net loop gain as shown is H − F dB , while F
is greater than H the loop gain is negative, i.e. there is a loss.
It is when F is less than H that the loop becomes unstable,
the hearing aid whistles and is unusable. We have not
complicated this general idea with matters of phase, these
enter feedback discussions later in Chapter 6.

An offshoot of this difficulty which in fact also affects
public address systems is illustrated in Fig. 2.4(ii). We can
reasonably assume that the amplifier gain/frequency
characteristic is "flat" but because of their miniature sizes
the microphone or earphone may have a response with a
pronounced peak as shown by curve (i) ("response" will have
more meaning after Chapter 5). If the level of the loop gain
at which instability occurs (H > F) is as marked then the
system is unstable at frequency f , yet overall the gain is
well below that permissible. If the peak in the response can
be removed more amplifier gain can be used as shown by
curve (ii) and still with stability. An important design
requirement is therefore that the transducer response
characteristics should be smooth. Another is that stability
is gained and therefore greater amplification allowed, if the
attenuation (F) of the air-path from earphone to microphone
is kept high. If a standard circular cross-section plastic ear
fitting (Fig.2.5) is insufficient owing to acoustic leakage
between it and the auditory canal a plastic mould specially
made for the particular ear is used. This provides a better
seal and effectively creates an acoustic disconnexion or at
least a very high attenuation in the feedback path.

2.3.2 Air-Conduction Aids

Fig.2.5 illustrates a popular "behind-the-ear" type of aid.
The amplified sound passes from the earphone through a
small plastic tube into the ear, hence the description "air-
conduction". The microphone may be situated at top or
bottom of the case facing forward, the integrated circuit
amplifier and battery are also contained within the case.
The whole unit "sits" on and behind the ear with the ear
fitting pushed into the ear canal. There are many variations,
for example (i) with the microphone, amplifier and battery

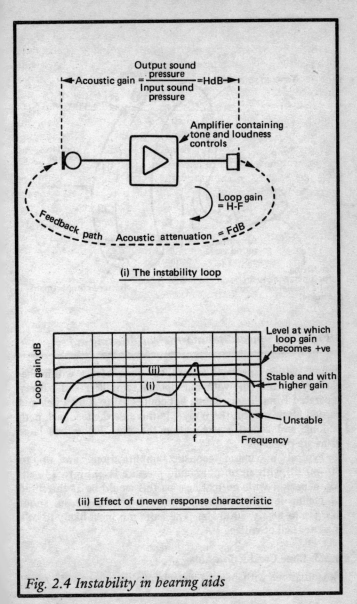

Acoustic gain = $\dfrac{\text{Output sound pressure}}{\text{Input sound pressure}}$ = HdB

Amplifier containing tone and loudness controls

Loop gain = H−F

Feedback path   Acoustic attenuation = FdB

(i) The instability loop

Loop gain,dB

Level at which loop gain becomes +ve

(ii)

(i)

Stable and with higher gain

Unstable

f   Frequency

(ii) Effect of uneven response characteristic

Fig. 2.4 Instability in hearing aids

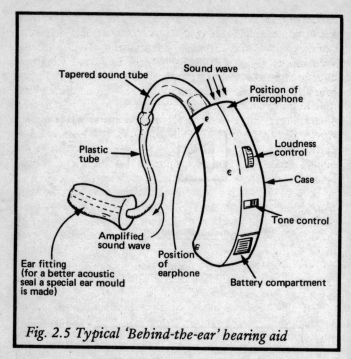

Fig. 2.5 Typical 'Behind-the-ear' hearing aid

housed within spectacle frames, (ii) completely "within-the-ear" types and (iii) with microphone, amplifier and battery separated from the earphone but connected to it by a flex and carried elsewhere on the body. In this case the feedback path is long and acoustic attenuation therefore high, thus large gains are possible.

Typical maximum acoustic amplifications are up to 60–80 dB with an output sound pressure level up to 130 dB (on a person with normal hearing this would be at threshold of feeling level). Frequency range is generally from about 100 Hz to nearly 5000 Hz. The battery is most likely to be a mercury cell of voltage 1.4.

2.3.3 Bone Conduction Aids

Not all those with hearing defects can be helped with an air conduction aid because for some the mechanical system from ear drum to middle ear (Fig.2.1) may be defunct. In such cases, provided that the inner ear mechanism is functioning satisfactorily, it is possible to inject the sound vibrations directly from the *mastoid* bone. This bone can be felt just below the skin behind the ear and one of the cavities within the mastoid bone structure connects with the middle ear. Sound vibrations however, reach the auditory nerves mainly via the bony structures of the head. In fact we hear ourselves partly through bone conduction when we talk, hence we seldom agree that tape recordings of ourselves are realistic because these rely on air conducted sounds only.

Because the sound vibrations are injected into a solid structure rather than into the air a special kind of *bone conduction receiver* is used. It may be held in position on the mastoid bone by means of a spring headband or even fitted to the end of spectacle frames. In a way it "hammers" on the mastoid instead of vibrating a diaphragm and typically the variable-reluctance principle is employed (discussed later in Sect.5.3).

Overall therefore, unless in the extreme case a person suffers from *nerve deafness*, that is, the failure of the auditory nerves in the inner ear to transmit to the brain, electronics is of untold benefit.

CHAPTER 3. THINGS WE HEAR

Life is such today that the difficulty is often in *not* hearing sounds, wanted and unwanted ones assail us from all directions. In this Chapter we examine the technical facets of a few basic sounds the most important of which surely is the voice, the other half of our acoustic communication chain.

3.1   THE VOICE

We have studied hearing in the previous chapter and undoubtedly the inner human mechanism is most complicated. That of the voice is no different, we take the ability of speaking for granted yet when we do so we are not only generating sounds but modifying them by the oddest of artifices, shifting tongue, lips, palate and larynx and even blowing through our teeth, all of which is controlled, rapidly and with apparently no effort, by the brain. Yet each of us has differences such that among thousands of people one particular voice is easily recognized. As with hearing, we look at the mechanism first, not only is this of interest but it is useful in the study of speech synthesis for electronic speech production.

3.1.1 Human Mechanism

Fig.3.1 is a sketch of the voice organs. They are contained within the *vocal tract* comprising throat, mouth and lips. Through the neck run two separate ducts, the *oesophagus* and the *trachea*. The latter is the important one here, it is some 10–13 cm long and it is the air passage containing the *larynx* (or voice box). The oesophagus is the channel for food and the *epiglottis* is an ingenious valve arrangement which normally allows air to pass through the trachea but closes this passage when food or liquid is being swallowed. The larynx contains the *vocal cords*, two bands of tissue positioned behind that part of the throat commonly known as the Adam's Apple. These tissues or membranes are elastic

47

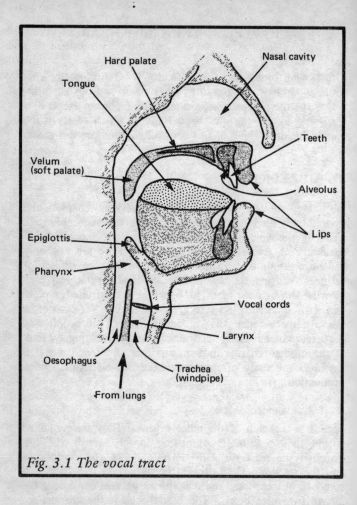

Fig. 3.1 The vocal tract

and are arranged so they vibrate or "buzz" when air is forced up between them, rather like the reed vibrates in a clarinet. Air then passes up the vocal tract in a series of rapid pulses, the range is about 100–200 per second for men and 150–300 or more for women, these figures are averages and there are obviously large deviations. Frequency is controlled by muscles which change the tension in the cords and loudness is controlled by the pressure put on the air stream through them. The vocal tract is effectively a tube, open at the mouth and as such acts as an acoustic resonator with more than one resonant frequency. (We will look at the technicalities of resonance in air columns in Sect.3.2.1 and of wind instruments, as is in fact the vocal tract, in Sect.3.2.1.1, the reader may perhaps wish to read these two sections before continuing here.) These resonant frequencies are known as *formants* and they change in frequency while we move tongue and lips in the process of joining the elementary speech sounds together, more technically known as *articulation*.

Talking is natural and easy, so much so that it is difficult to realise just how much tongue, lips and throat move around in the process. Thus speech analysis (taking apart) and speech synthesis (putting together) are quite complex for there are many different individual *sounds* involved. There is no point in our looking at all of these, one or two typical ones from each group ensures a basic understanding of the process.

3.1.2 Speech Analysis

We start with the vowel sounds, not just the five a, e, i, o and u, but over twenty for not only is there more than one sound to each (e.g. b*o*g and b*o*rn, b*i*t and b*i*rd) but the basic vowel sounds also appear in combinations known as *dipthongs* (e.g. h*ea*r, b*oi*l). The different vowel sounds arise from excitation of the vocal cords (voicing) and shaping the vocal tract in certain ways to change the acoustic resonances which have been described and labelled above as formants. The mouth is opened to various degrees. The formants are usually designated F_1, F_2, F_3, etc., F_1 being lowest in frequency and of greatest amplitude. Typical formant amplitude/frequency diagrams are shown in Fig.3.2 for the a in at and the i in it spoken by one particular male voice.

Consonant sounds use additional forms of excitation with or without generation by the vocal cords, they are subdivided as follows:-

(i) *plosives* are formed by blocking the vocal tract so that no air flows and then suddenly removing the obstruction to

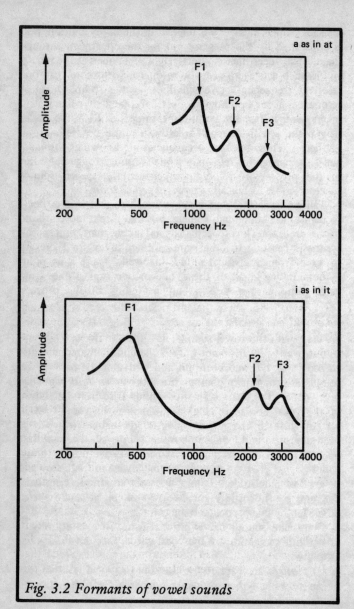

Fig. 3.2 Formants of vowel sounds

produce a puff of air. Lips, teeth and velum are active in this process. Two plosives in which the lips play the major part are p and b. When p is used in the word pit it is unvoiced (i.e. the vocal cords are not in use), when b is used in the word bit it is voiced. T and d are formed at the upper teeth, g and k at the velum.

(ii) *fricatives* involve the lower lip and teeth with air blowing through as when f is used in fill (voiced) or v in van (voiced). With th as in both (unvoiced) and in then (voiced), the lower lip takes no part.

(iii) *sibilant fricatives* are the hissing sounds. Air is blown over the cutting edges of the front teeth with the lips slightly parted. An unvoiced sibilant fricative is s as in this , a voiced one is z as in zone .

(iv) *semi-vowels* are so classed because they are relatively open sounds as are vowels. They are formed like vowels and are therefore all voiced but formed via added constrictions made by lips or tongue. In this class are w as in wet , l as in let , r as in rat and y as in you .

(v) *nasals* — for these the soft palate is lowered so blocking the mouth and forcing the air stream to pass out via the nose. Nasal sounds are all voiced, m for example as in make is sounded with the lips closed, on the other hand the lips are slightly open for n as in not and the tongue is raised into contact with the forward part of the hard palate or alveolus.

We should now have little difficulty in placing voice sounds in their correct categories and also have some idea of the facial and vocal tract movements, also whether the vocal cords vibrate or not. In *phonetic analysis* vocal sounds are known as *phonemes* and each is represented by a symbol. There are well over 40 different sounds so many symbols are needed, usually obtained from the English and Greek alphabets together or alternatively for computer applications by using singles and pairs of English letters, for example ϵ in the first case represents the phoneme e in bet , in the second case it is EH . Thus speech can be expressed in writing not only by the words we know but also in terms of the actual sounds, that is, in phonemes. This latter method has application in design and assessment of speech carrying systems and in the generation of speech by electronic methods.

3.1.3 Speech Synthesis

Following on from the idea of analysing speech into its constituent elementary sounds, it would appear that if electrical analogues of these can be held as recordings, or stored digitally in a computer memory[(4/3.3)] or even generated, then brought into use serially as required, we have the elements of electronic speech generation. Alternatively words themselves may be recorded by a talker and recalled as required for assembling into phrases under computer control. However this is hardly synthesis and it has the major disadvantage of the requirement of a large amount of memory. True speech synthesis techniques occupy much less memory but are undoubtedly more complex.

The preceding section shows that speech excitation needs two separate generators, one a variable oscillator rich in harmonics, (i.e. tending towards a square-wave output) to produce the formant frequencies and the other a noise generator to simulate air turbulence or hiss. These are in Fig.3.3 which shows diagrammatically the components of a *basic* speech synthesizer, i.e. with no frills. All the units with control lines terminating on them are *voltage controlled* either in frequency or amplitude. (We will be studying voltage control in Chapter 8.) $F_1 - F_3$ is a set of formant filters which provide the three formants as shown typically in Fig.3.2. The nasal formant frequency is supplied directly from the main oscillator and the white noise generator[(3/3.2.2.2)] supplies both the fricative formant unit and $F_1 - F_3$ via amplifiers with voltage controlled gain. (Some readers may find difficulty in getting to grips with Fig.3.3, if so it may be helpful to read this section again after studying Sect.8.2.)

Single or double letter symbols to represent the phonemes (phonetic text) are entered into the computer memory via the normal keyboard. Suppose we wish to hear the word "synthetic" spoken. A particular phonetic text input might be as follows:-

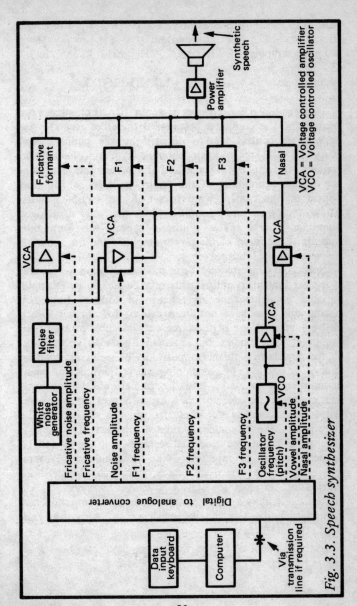

Fig. 3.3. Speech synthesizer

VCA = Voltage controlled amplifier
VCO = Voltage controlled oscillator

Synthetic speech

Power amplifier

Fricative formant

F1

F2

F3

Nasal

VCA

VCA

VCA

VCA

White noise generator

Noise filter

VCO

Digital to analogue converter

Data input keyboard

Computer

Via transmission line if required

Fricative noise amplitude

Fricative frequency

Noise amplitude

F1 frequency

F2 frequency

F3 frequency

Oscillator frequency (pitch)

Vowel amplitude

Nasal amplitude

53

```
word           — S   Y   N   T   H   E   T   I   C
                 |   |   |   \  /    |   |   |   |
phonetic symbols — S  IX   N   TH  EH   T  IX   K

phonetic text  —              SIXNTHEHTIXK
```

Conversational speech runs at about 10–20 phonemes per second and the computer programme therefore generates the appropriate control bits as directed by the phonetic text every 50–100 ms. Each phonetic symbol (one or two characters) is taken in turn and the programme computes the digital equivalents of the control voltages which must appear on the control wires to set all the 9 variable units. These voltages are supplied by the digital/analogue converter.[4/4.8.1] The synthetic speech waveform thus appears at the input of the power amplifier and the speech is delivered by the loudspeaker.

Speech so generated can be of moderately good quality but it is clear that many refinements may be added, for example, variable phoneme time and change of formant frequency during this time have not been catered for nor has speech delivery rate. None of these presents insoluble problems for computer programmes but naturally adds to complexity.

We now have a further opportunity of seeing *information theory* in action by examining how transmission plant economies can be made. Looking again at Fig.3.3 it is evident that voltages are set up on the synthesizer control wires within microseconds yet thereafter do not change for the duration of the phoneme, a time of at least 50 ms. It is the realization of this which led researchers to the idea that a speech channel of say 300–3400 Hz could carry several conversations at once and much work has already been carried out on what are known as *speech compression* techniques. Firstly we recall Shannon's basic information rate formula relating the channel capacity c with the bandwidth W ,[5/1.5.4] i.e.

$$C = W \log_2 \left(1 + \frac{S}{N}\right) \text{ bits/s}$$

where $\frac{S}{N}$ is the signal/noise ratio. Then considering a 3100 Hz bandwidth channel having a 30 dB signal/noise power ratio (1000:1, ample because errors in speech systems are nowhere near as disastrous as they are with computer data), $C = 3100 \log_2 (1 + 1000) = 3100 \times 9.97$, i.e. more than 30,000 bits/s. Let us suppose that in our model synthesizer as many as 8 bits are needed per control wire into the digital/analogue converter[4/4.8.1] for it to determine the control voltage, i.e. $9 \times 8 = 72$ bits altogether. To generate speech therefore 72 bits are required every 50 ms, i.e. a bit rate of nearly 1500 bits/s. Thus on this reasoning a single telephony channel theoretically can accommodate $\frac{30,000}{1,500} = 20$ separate talkers, equivalent to 10 bothway conversations. *Information content* calculations[5/1.5.2] which allow for the *redundancy* in language (we can make ourselves understood just as well if a lot is left out) show that theoretically only about 5 bits per phoneme are essential. This leads to the conclusion that many more conversations could be carried *with redundancy* than our own calculations indicate. There is no single practical answer however, so much depends on the system and overall speech quality provided.

To transmit speech over such a system, instead of a computer reading phonetic text there would be an analyser "reading" a microphone, in both cases the output is the same, a string of binary digits to control the 9 variable units of the synthesizer. Thus at the far end of the line the bits are received by the digital/analogue converter and when several systems share the same transmission line the latter is allocated to each system in turn as with p.c.m. channels.[5/5.5.3.1]

In this exercise on information rates we have looked at the elements of our own idea of a speech compression system as a help in understanding the problems. Experimental ones with the same intent of economizing in bandwidth are sometimes known as *vocoders* (voice coding). On the practical side it must be admitted that such systems have been under development for several decades, the gains in transmission efficiency are realizable but their complexities are such that optical fibre transmission[5/7] with its relatively cheap bandwidth may very well make further development unlikely. On the other hand computer generated speech has a host of applications.

3.1.4 Speech Characteristics

Most of us are used to seeing graphs of sine waves, square waves and those of equally regular form. Speech waveforms however are transitory, oscilloscope portrayals are lively but irregular and orderless consisting as they do of so many different frequencies at once, all of varying amplitude. The speech waveform is therefore difficult to measure and instantaneous values have little meaning, thus we shall find that most measurements have to be averages taken over a period of time, sometimes as much as several minutes.

3.1.4.1 Power Levels

Generally, vowels contribute more of the power in speech than do consonants yet peculiarly enough consonants play a greater part in speech intelligibility. The actual power generated by a person talking at a "normal" level expressed as the average over a sufficiently long period of time for it to remain constant (known as the *long term mean power*) is about 10 μW and of interest is the fact that if only vowels are spoken this power rises to 20–30 μW. On shouting the acoustic power may rise to 1 mW and if talking in a whisper it can fall to as low as 0.001 μW, for the vocal cords are then not in use. Speech therefore has a dynamic range of about

$$10 \log \frac{10^{-3}}{10^{-9}} = 60 \text{ dB}.$$

3.1.4.2 Peak Factor

One of the peculiarities of a speech waveform is the very large fluctuation of instantaneous level, this is called the *peak factor*, expressed more precisely as the ratio of the peak to r.m.s. pressure. The peak factor of a sine wave is $\sqrt{2}$, (2/1.3.3) so the peak voltage is about 3 dB (20 $\log\sqrt{2}$) greater than the r.m.s. voltage. With speech, calculation is not possible, measurements have to be made instead and that for the r.m.s. is taken over a known period of time. Simultaneously observations are made of the peak values reached. It has been determined that if the period of time is short, say a small fraction of one second, then the peak values may rise on average some 10 dB above the averaged r.m.s. value while for a longer period of several seconds this increases to some 20 dB. Such high peak factors are of concern in amplifier design. If the long term r.m.s. sound pressure produced by an amplifier and loudspeaker system carrying pure tones only requires, say, 1 W of output power, then to reproduce faithfully the peaks of the wave the amplifier must be capable of 2 W output (3 dB above 1 W) without distortion. With speech instead of pure tones things get a little out of hand if we aim for perfection for the amplifier must now be capable of 100 W output (20 dB above 1 W) if distortion is to be avoided entirely. In practice some relaxation is usually made based on the fact that the very highest peaks in the speech waveform occur relatively infrequently.

Very approximately, for an amplifier carrying continuous speech, peak distortion occurs for 10% of the time if the maximum peak factor it can handle is 6 dB but only for 1% of the time for a peak factor of 12 dB and 0.01% for a peak factor of 20 dB. If we class distortion for 0.01% of the time as perfection, as discovered above, our amplifier needs a maximum power output of 100 W, however by reasonably allowing the distortion time to rise to 1% the maximum power output need only be about 16 W and with this it is still unlikely that the distortion would be noticeable.

3.1.4.3 Measurements

Intelligibility of speech is determined by *articulation* tests. Sounds or syllables, some of which may be meaningless are spoken over a system in which various parameters may be adjusted such as frequency range, loudness or speech clipping (as in preceding Section). Several observers write down what they think was heard. The degree of success naturally would depend also on the speech clarity of the talker so the tests are arranged to eliminate this effect. The *articulation index* is the percentage of sounds correctly received. As an example of such tests it has been determined that removal of the lowest frequencies (below 500 Hz) does not greatly affect speech intelligibility but removal of the higher ones (above 2500 Hz) does affect it because about 20% of the sounds are incorrectly heard (80% articulation index). This is in agreement with what we have already found in Sect.3.1.4.1, that consonants, which contribute more to the higher frequencies, play the greater part in speech intelligibility.

Electrically, measurement of the speech waveform is usually as a voltage, even though the intention may be to refer to the power. Because of the irregularity of the peak factor, with conventional instruments from neither a measurement of the power nor of the voltage can we derive figures with any consistency of meaning. Generally therefore the speech waveform is measured on a voltmeter with specified *rise time* (the time required to reach 99% of the final steady reading). This is important because with short duration inputs the degree of swing of a voltmeter needle depends on the rise time, shorter rise times giving greater deflexions.

The UK *speech voltmeter* has a rise time of 230 ms, that of the American VU (volume unit) meter[5/2.1.1.2] is 300 ms. These voltmeters are calibrated in decibels relative to a particular zero, in the first case 1 V, in the second 0.7746 V (equivalent to 1 mW in 600 Ω). Certain rules must apply for interpreting the meter indications because the pointer fluctuates rapidly, thus for example, the VU-meter is read over a 10 s period by excluding the two or three highest deflexions of the pointer and taking the average level of the remainder, the observer therefore is skilled.

Several other types of speech measuring instruments have been developed, especially automatic ones which obviate the need of a human observer.

3.2 MUSICAL INSTRUMENTS

Ever present is music, abounding in a myriad of different forms which reside in the numerous instruments handed down to us from long ago. Nowadays many musical instruments are copied and even extended by electronic methods, but we leave this for discussion in Chapter 8. In this Section we look at the underlying technical features on which the creation of music is based.

3.2.1 Vibration of Air Columns

Wind instruments (including church or "pipe" organs) produce their harmonious sounds through resonance of an air column. Columns are usually associated with vertical objects but for lack of a better one, here we understand the word to signify a volume of air enclosed within a tube whether the latter is held vertically or not. The everyday experience of running water from a tap into a kettle or bottle where the sound emitted rises in pitch as the vessel fills, sets the scene.

We looked at reflection generally in Sect.1.4.2 and used as a parallel the return of a signal current on completing its travel along a mismatched transmission line.[5/6.1] In this Section the transmission line bears a greater resemblance because the sound waves are bounded on all sides by the walls of a tube. Let us begin with a cylindrical tube closed at one end and with, say, a tuning fork held at the open end. Sound waves enter the tube and travel to the closed end where, unless completely absorbed, which is unlikely, some are reflected. The incident and reflected waves together add or subtract according to their relative phases and the net magnitude at any point along the tube can be calculated for any instant in time. We talk in terms of particle displacement (Sect.1.3.4) because it is thereby easier to understand our first supposition which is that displacement is zero at the closed end of the tube for clearly no particle can vibrate to and fro when it meets the solid obstruction of the closure. We could work with sound pressure instead but because displacement and pressure are out of phase with each other, it is less confusing if we choose one and stay with it. The point of zero displacement is called a *node* (a stopping point – from Latin, nodus, a knot).

Consider a sound wave compression at its maximum level at the open end of the tube, displacement is maximum because there is no obstruction and this is an *antinode*. The wave travels down the tube and if arrangements can be made for it to be reflected from the closed end and arrive back at the open end so that it reinforces the wave existing there, the condition of resonance is set up. We can use some practical figures in conjunction with Fig.3.4(i) to get to grips with this. Consider a wave of frequency 100 Hz. Its period[2/1.2.2] is therefore

$$\frac{1}{f} = 10 \text{ ms} \text{ and wavelength } \lambda = \frac{c}{f} = 3.44 \text{ m. The graph on}$$

the left of the Figure represents the sound wave entering the tube and at t = 0 it is a compression. By making the tube length ℓ = 0.86 m which is one-quarter of a wavelength, we see that this compression reaches the closed end in a time

$$t = \frac{\ell}{c} = \frac{0.86}{344} \text{ s} = 2.5 \text{ ms} . \text{ On reflection}[5/6.1.4] \text{ (which in}$$

a very simplified way we might consider as the particles bouncing back off the end wall) the wave travels back and arrives at the open end 2.5 ms later, that is, at t = 5.0 ms , so travelling a total of half a wavelength. At this time the incoming wave itself has moved through 180° so because of the phase-change on reflection, both incoming and reflected waves are in phase, the condition required for resonance. Then

$$\lambda = 4\ell \text{ and since } f = \frac{c}{\lambda} , \text{then } f = \frac{c}{4\ell}$$

showing that the frequency of resonance is inversely propor-

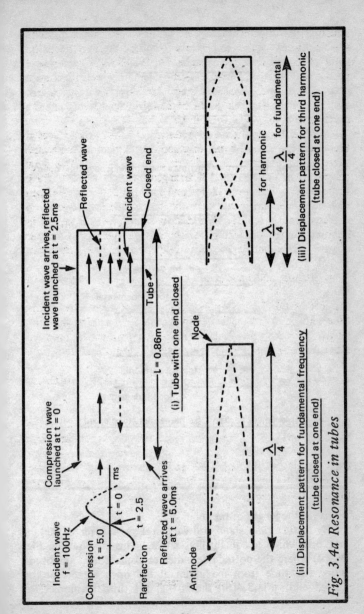

Fig. 3.4a Resonance in tubes

(i) Tube with one end closed

Incident wave f = 100Hz

Compression wave launched at t = 0

Incident wave arrives, reflected wave launched at t = 2.5ms

Reflected wave arrives at t = 5.0ms

Reflected wave

Incident wave

Closed end

Tube

L = 0.86m

Compression

Rarefaction

t = 5.0 ms

t = 2.5

t = 0

(ii) Displacement pattern for fundamental frequency (tube closed at one end)

Antinode

Node

$\frac{\lambda}{4}$

(iii) Displacement pattern for third harmonic (tube closed at one end)

$\frac{\lambda}{4}$ for harmonic

$\frac{\lambda}{4}$ for fundamental

61.

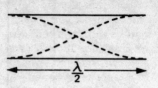

(iv) Displacement pattern for fundamental frequency (tube open)

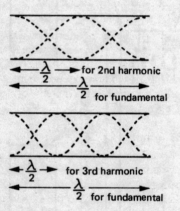

$\frac{\lambda}{2}$ for 2nd harmonic

$\frac{\lambda}{2}$ for fundamental

$\frac{\lambda}{2}$ for 3rd harmonic

$\frac{\lambda}{2}$ for fundamental

(v) Displacement patterns for harmonics (tube open)

Fig. 3.4b

tional to the length of the tube.

By letting time stand still we can draw a different picture but one which is equally illustrative. This is shown in Fig. 3.4(ii) where the dotted line indicates displacement amplitude along the tube from maximum at the open to zero at the closed end. Fig.3.4(iii) shows that the third harmonic also meets the requirements of an antinode at the open and node at the closed end, higher *odd* harmonics do likewise. It is evident too that even harmonics are not reinforced because they cannot meet the resonance conditions.

Moving next to tubes with both ends open and appreciating that there must be antinodes at both open ends, Fig.3.4(iv) shows that the length of tube for resonance must be half a wavelength. There is some reflection at both ends because of the change in boundary conditions, from the enclosure of the tube to the freedom of the open air; somewhat akin to a transmission line of one characteristic impedance meeting another of different impedance.(5/6.1.4)   In (v) of the Figure is demonstrated the condition of both even and odd harmonics being supported.

This is theoretical, the inconsistency with what occurs in practice arises at the open ends where the displacement pattern does not finish neatly as is suggested in the Figure. In fact the vibrating air column extends slightly outside an open end, so modifying the actual length of tube required for resonance at a given frequency.

3.2.1.1 Wind Instruments

In the preceding Section we first considered an air column set in vibration by a tuning fork. Woodwind instruments similarly need a pulsating air stream and this is produced by blowing across a column of air as in the flute or forcing air from the mouth through a reed which vibrates mechanically as would a tuning fork, this is the method used in a clarinet. With brass instruments such as the trumpet, the player's lips vibrate as air passes them into the instrument.

Different notes are obtained by changing the length of the air column, the various methods are soon evident on watching a particular instrument being played.   The church organ differs in that it uses a separate tube for each note instead of a single one with the air column varied in length. We see too how instruments may differ fundamentally in their sounds by being closed at one end (e.g. flute) with the characteristic tone from the accentuation of the odd harmonics, or open (e.g. clarinet, trumpet) where the even harmonics also join in.

3.2.2 Vibration of Strings

The principles discussed in the Section above regarding waves travelling along a tube and their subsequent reflection at a discontinuity applies equally to waves in a *string*. By string is inferred a length of catgut, cord or wire stretched between two supports and highly elastic so that any deformation by plucking or bowing is quickly restored.

If a string under tension is plucked at one end a wave travels along it to the far end clamp. The wave velocity is determined by the tension in the string, T and its mass per unit length, m , related by the formula:

$$\text{Velocity (V) m/s} = \sqrt{\frac{T}{m}}$$

where T is in newtons and m in kg/m. Although the wave progresses along the string the actual string movement is perpendicular to this direction. The clamped ends are nodes for there is no string vibration at these points, thus the parallel with the air column in a tube is not exact for a tube must always have an antinode at one end. The simplest mode of string vibration therefore has a single antinode at the centre of the string as shown in Fig.3.5(i) which indicates particle displacement. In (ii) and (iii) of the Figure both even and odd harmonics arise and the distance between any two successive nodes is $\frac{\lambda}{2}$. Let ℓ represent the length of the string and for the fundamental, $\ell = \frac{\lambda}{2}$ $\therefore \lambda = 2\ell$

$$\therefore \quad f = \frac{1}{2\ell} \sqrt{\frac{T}{m}} \quad \text{or} \quad \frac{\sqrt{T}}{2\ell\sqrt{m}}$$

that is, the fundamental note emitted is (i) inversely proportional to both the length of the string and the square root of its mass per unit length, (ii) directly proportional to the square root of the tension.

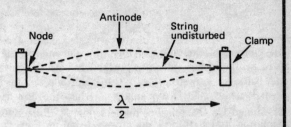

(i) Vibration at resonant frequency (fundamental)

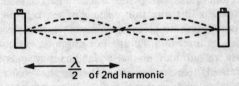

(ii) Vibration at frequency of 2nd harmonic

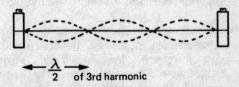

(iii) Vibration at frequency of 3rd harmonic

Fig. 3.5 Resonance in strings

From (ii) we see how pianos, violins and similarly functioning musical instruments are tuned, that is, by adjustment of string tension.

3.2.2.1 String Instruments

The method of excitation of the string depends on the instrument being played. It can be plucked, as with the guitar, harp and banjo or *bowed* as with the violin and cello. Bowing perhaps needs a little explanation. When the bow is drawn across a string the latter is pulled transversely so continually increasing its length and therefore tension until eventually it suddenly frees itself from the frictional grip of the bow and slips back. It is again caught by the bow and the process repeats, the string is thus forced into vibration.

3.2.3 Percussion

Nowadays we seem to beat upon almost anything to get music out of it, from drums, xylophone and castanets to steel barrels and dustbin lids. As with the plucking of a string, the generation of sound waves by percussion seems visually obvious but in fact analysis is more difficult. For example, so far we have consistently talked in terms of a fundamental plus its harmonics. With percussion it is better to change from "harmonics" to *overtones* which is a term used in a wider sense. It describes all higher frequencies associated with a fundamental but these not necessarily being a whole number multiple as is a harmonic. In other words, overtones include not only harmonics (2nd, 3rd, 4th, etc.) but also other ratios (for example, 2.7, 5.4, etc.).

The tuning fork is a percussion instrument and is one example of a bar clamped at one end. It vibrates transversely. At its fundamental frequency the clamped end is obviously at a node and the free end at an antinode. The frequency is determined not only by the length of the bar but also by the elasticity and density of the material and by the *radius of gyration*. The latter is a term met in applied mechanics and it handles the fact that the bars may have any shape of cross-section, even hollow. It is quoted here so that we may be convinced of the complexity of calculations with regard to percussion instruments and leave well alone. Nevertheless, what we have already discussed, especially with regard to strings will help us to appreciate what follows.

When struck, the clamped bar produces overtones, the first being over six times the fundamental and therefore relatively weak as would be a 6th or 7th harmonic.[2/1.4] The tone of a bar which is not clamped at one end but simply supported at two points as for a single xylophone note is richer in overtones, the first being less than three times the fundamental and therefore more lively.

Things get really complicated when we consider membranes stretched over the end of a tube or cylinder as in the case of a drum. At resonance at least we know that a node exists around the periphery with an antinode at the centre but because, unlike the air column or string, the membrane vibrates in more than one direction, our analysis stops here. However, from observation of the "big bass drum" it is clear that as the diameter increases the resonant frequency falls and when the tension is increased the frequency rises.

Also defying simple analysis are gongs and circular plates such as bells and cymbals. These are individually shaped and their overtones are not harmonic. The fundamental frequencies of xylophones, bells, etc. are well defined so that these instruments play musical notes. Those of the drum or cymbals are not so definite, hence these instruments cannot be used for the melody, they give backing only.

3.3 STEROPHONIC SOUND

Realism is added to sound reproduction by the technique generally known as *stereo*, short for *stereophonic sound*. Stereo comes from the Greek, meaning "solid", literally therefore stereophonic refers to "solid sound", perhaps better translated as *three-dimensional sound*. Its aim is to further the illustration of actually being present and listening to the originating sound source. We have two ears and between them and the brain the location of a sound in the horizontal plane is sorted out. Therefore when confronted by a *mono-phonic* system (mono = one) in which all sounds arrive via a single microphone, the spatial effect is lost and we are no longer aware of sounds arriving from different places. Stereophony overcomes this to a certain extent by the use of two or more microphones followed by completely separate channels whether direct or recorded, feeding individual loudspeakers. Even though the loudspeakers may be relatively close together in the home, the sound appears to be "spread out" which is especially advantageous in the reproduction of large orchestras where a single microphone destroys the feeling of size. With stereo the resemblance of movements of an originating sound source is also impressive. Thus adding as it does a third dimension and giving greater perspective and clarity, 2-channel stereo can be said to have advanced hi-fi one step further, while systems with more channels add even more.

3.3.1 Judging Sound Direction

For sounds not directly in front of us there is a difference in arrival time (and therefore phase) at the two ears, also a difference in intensity. It is these differences which are fed to the brain and from which we get the sense of sound direction. To see this in more detail and accordingly to appreciate what information the brain is denied with mono but partly provided by stereo, let us look at the elements of the process as illustrated in Fig.3.6. A single source emits sound waves which are heard by a listener as shown. Evidently each sound wave reaches the right ear first and the left ear later by a time dependent on the distance d , that is

$$t = \frac{d}{344} \text{ secs}$$ where d is in metres. If for example

d = 10 cm then t ≈ 0.3 ms and for example at 1000 Hz which has a period of 1 ms, the phase difference at the two ears would be 0.3 x 360 = 108°.

In our example the head constitutes an obstruction to sound reaching the left ear but not that reaching the right ear. Sect.1.4.4 shows that not only does this cause some general attenuation but also that the higher frequencies are attenu-

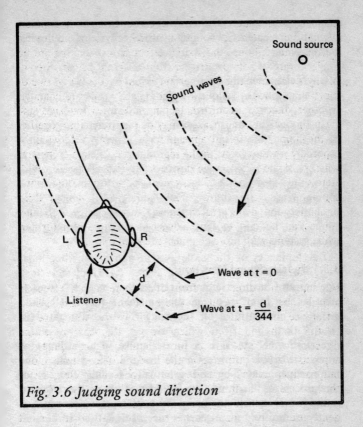

Fig. 3.6 Judging sound direction

ated more than the lower ones at the left ear. Thus the brain receives information from the ears on the phase difference, the amplitude difference and how the latter changes with frequency. We ourselves are not even conscious of this unbelievably complex process but from it we derive a fairly accurate sense of the direction from which a sound is arriving, notwithstanding any reflexions which may tend to cloud the issue.

The natural conclusion to this is to use a dummy head fitted with microphones at the ear positions for sound pick up and terminating in two earphones at the remote end of the channel. This *binaural* system was earlier brought to fruition but naturally could not compete with later systems employing loudspeakers.

3.3.2 Multi-Channel Stereo

Two-channel stereo arose from the idea of placing two microphones in different positions and locating the two loudspeakers the same distance apart as in Fig.3.7(i). In this way the listener's brain is fed with much of the phase and amplitude information necessary for a realistic effect. A host of conditions may upset the acoustic balance between the two loudspeaker sound outputs, especially listener's position and reflexions, therefore a balance control is added. By moving the slider up in the Figure the R-side is more heavily shunted (so reducing the R loudspeaker output) and the L-side less and vice versa.

Developments of the basic system above use directional microphones (discussed later in Sect.5.1.1.4) placed close together and having maximum response at, say, 90° to each other so that in effect one picks up mainly from the left while the other concentrates on the right. This is illustrated in Fig.3.7(ii).

Four-channel stereo is a further move in an attempt to provide *surround* or *wrap around* sound. In essence four microphones may be used eventually feeding four loudspeakers placed around the listener, this is the *discrete* method. However, the need of four separate channels is rather prohibitive, thus systems have been derived to be able to create the same illusion but using two separate channels only, the *matrixed* method.

3.4 NOISE

Some of the sounds we hear we call *noise*, the ones which are usually the least welcome. In this Section we firstly consider acoustic or ambient noise and there is certainly no need to describe something we all experience incessantly. *Why* we find one sound pleasant and not another, Nature has not yet

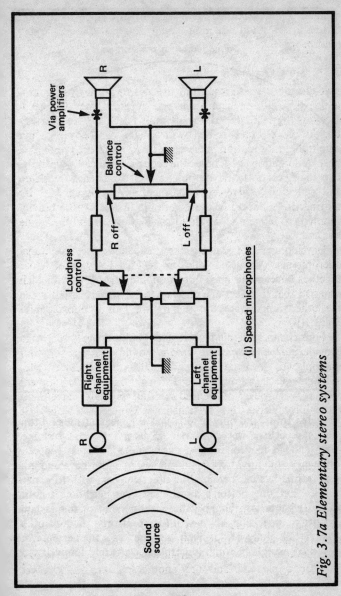

Fig. 3.7a Elementary stereo systems

(i) Spaced microphones

R
L
Via power
amplifiers
Balance
control
R off
L off
Loudness
control
Right
channel
equipment
Left
channel
equipment
R
L
Sound
source

71

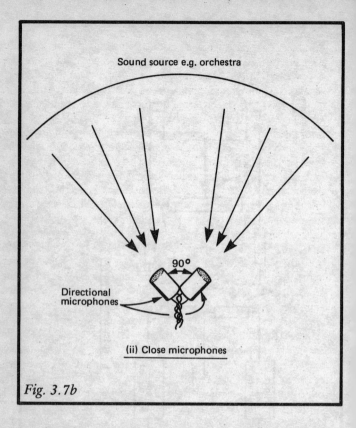

Sound source e.g. orchestra

90°

Directional microphones

(ii) Close microphones

Fig. 3.7b

revealed to us, the only clue is that the pleasant musical tone has a repetitive waveform rich in harmonics and sustained long enough to produce a musical "note" whereas generally the bangs, thumps, hisses and crashes of noise have sporadic waveforms lacking perhaps the order the ears wish to experience. Yet one person's music may be a neighbour's noise. What is important though is the measurement of this variable quantity especially in view of our efforts to control it. People talk glibly of "decibels of noise" as if the decibel were invented especially for this purpose, we surely know better.

3.4.1 Measurement

Audible noise measurement is on the basis of sound pressure level as defined in Sect.1.3.1. The instrument used is called a *sound-level meter*. Because the term "sound pressure level" implies the known reference level of 2×10^{-5} N/m^2 (Fig.1.7), we can talk directly in terms of decibels without quoting the reference level each time. For a simple "one-off" measurement a hand-held sound-level meter is adequate but for more analytical work a calibrated magnetic tape recorder may be used so that the sound sample can be repeated to an oscilloscope, graphic level recorder (draws the waveform envelope on moving graph paper) and/or frequency spectrum analyser (measures the noise level in different frequency bands). A simplified diagram of a sound-level meter is given in Fig. 3.8(i). Ignoring the "weighting networks" the system must be "flat" in that, irrespective of frequency, equal sound pressures at the microphone give rise to the same meter deflexions. The range of sound pressure levels is large hence an attenuator follows the microphone so that the microphone amplifier is not overloaded at high noise levels. The attenuators enable the range to be changed in, say, 10 dB steps, single figures being indicated on the meter scale. The *averaging* circuit prevents the pointer trying to follow peaks and averages over 0.2 or 1 second as required. The longer time is especially useful for high-energy, short duration, relatively infrequent noise peaks (the bangs and clicks).

The *weighting networks* give predetermined "weights" or compensation to certain frequencies. Taking the commonly used A network as an example, this simply creates a loss at the lower frequencies which increases as frequency falls so needing a higher s.p.l. for the same meter reading compared with that for the higher frequencies. The net effect on the sound-level meter response is indicated in Fig.3.8(ii) and this is necessary because we are "looking" at a sound pressure level, not listening to it. In listening our ears tell a different story because their response is not "flat" with frequency but less sensitive at the lower ones, especially when the sound pressure is low. The A-weighting network therefore simulates average hearing by gradually

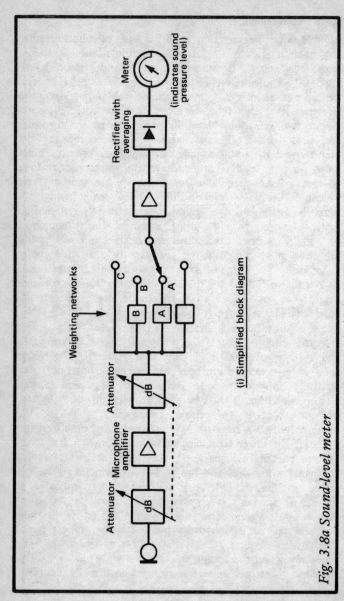

Fig. 3.8a Sound-level meter

(i) Simplified block diagram

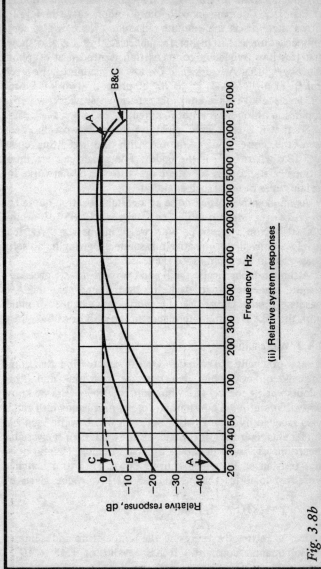

(ii) Relative system responses

Fig. 3.8b

introducing some 50 dB loss as frequency falls from 1000 Hz to 20 Hz. Comparison with the 40 phon curve in Fig.2.2 shows that this is the amount required. There is also some compensation at the higher frequencies. Fig.2.2 also shows that less low frequency compensation is needed at 80 phon, hence weighting A is designed for low to medium noise levels and B for higher levels up to 70–80 phon. Network C is used for levels above this and for experimental work where a simulation of loudness is not required.

So that it is clear as to which weighting network has been used, the noise s.p.l. is quoted with the weighting code, e.g. dBx where x is the code. The weightings are internationally standardized. There are also special networks for certain types of noise such as aircraft.

As an example, office noise at a certain location quoted as 65 dBA [or dB(A)] infers a mean s.p.l. 65 dB above 2×10^{-5} N/m^2, measured with weighting network A. It is merely a convenient objective measurement giving us a useful assessment of the loudness of the noise.

Although theoretically the B and C weightings are necessary for the higher noise levels it has been found that the A is reasonably satisfactory over the whole range hence for much work the dBA is used in preference to the dBB or dBC.

3.4.2 White and Pink

It may originally have come as a surprise to find that some noise is coloured white, but now we have pink and other colours yet to come. The white variety,(3/3.2.2.2) as we know is *random* and has a constant power per unit bandwidth and it is so labelled by analogy with optics in which white light has similar characteristics, it contains all the colours of the variable spectrum. Thermal agitation produces random (white) noise in a resistance R which delivers power (P_n) to a matched noise-free load [as shown in Fig.3.9(i)] of value given by

$$P_n = kT(f_2 - f_1) \text{ watts}$$

where k relates the energy to the temperature and is known as Boltzmann's constant. It has a value of 1.38×10^{-23}

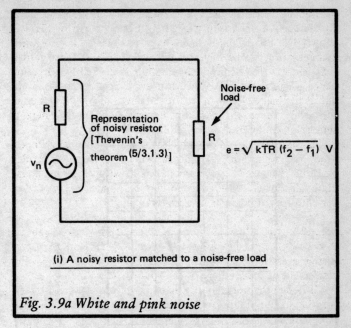

R

Representation
of noisy resistor
[Thevenin's
theorem [5/3.1.3]]

Noise-free
load

R

$e = \sqrt{kTR(f_2 - f_1)}$ V

v_n

(i) A noisy resistor matched to a noise-free load

Fig. 3.9a White and pink noise

joule/°K. T is the temperature of the resistor in °K.
$(f_2 - f_1)$ is the bandwidth and we note that the actual
frequencies of operation can be anywhere in the spectrum,
that is, given the same temperature, the same noise power is
generated at low as at high frequencies over equal bandwidths.
The load is matched hence this is the maximum power which
can be delivered and although at first this may seem odd, it is
independent of the actual value of R .

The r.m.s. voltage across the load, e is

$$\sqrt{RP_n} = \sqrt{kTR(f_2 - f_1)}$$

and this must be the same across the noisy resistor. The
generator e.m.f. v_n must therefore be 2 x e , i.e.
$v_n = \sqrt{4kTR(f_2 - f_1)}$ volts. For a practical example
consider a 1 MΩ resistor at the input of an audio amplifier
of voltage gain 1000, effective over 20 kHz. At 20°C (293°K)

77

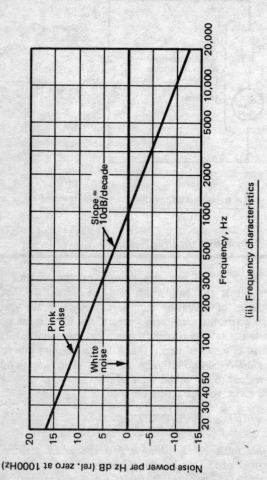

(ii) Frequency characteristics

Fig. 3.9b

$$v_n = \sqrt{4 \times 1.38 \times 10^{-23} \times 293 \times 10^6 \times 2 \times 10^4} \approx 18\mu V,$$

seemingly trivial but at the amplifier output it becomes 18mV, no longer negligible for if for example mixed with a signal at 1 V, the signal/noise ratio is only

$$20 \log \frac{1000}{18} \approx 35 \text{ dB} .$$

The thermal agitation noise voltage generated across a component is usually the basis of a discrete component noise generator. There are components which are more efficient from the noise point of view than resistors especially diodes and certain transistors, all that is required is to follow the particular component by an amplifier of high gain. I.C. noise generators are likely to employ methods more suited to integration techniques such as a shift register[4/4.4] controlled by a special clock.

The audio frequency spectrum of white noise is therefore the horizontal line on the graph in Fig.3.9(ii). Although the spectrum level is in terms of noise power per Hz it would not (and probably could not) be measured in this form but more likely in ⅓-octave bands, followed by calculation for 1 Hz. The characteristic represents the mean over a certain time, not the instantaneous conditions, for with random noise there might be no power at all within a certain small band at one instant, yet, it could rise to many times the mean value at another.

White noise, when reproduced by a loudspeaker, because of its high frequency content tends to sound rather "harsh". One of the more common variations, especially useful in musical note generation (see Chapter 8) is *pink* noise which has a falling frequency characteristic and to the ear is less harsh, perhaps best described as "rounded". The noise is a characteristic of semiconductors due to minute fluctuations in conductivity and is inversely proportional to frequency, hence the alternative description, $\dfrac{l}{f}$ *noise*. Its characteristic is also in Fig.3.9(ii), it has equal energy over bands of equal *ratio* $\left(\dfrac{f_2}{f_1}\right)$, for example there is the same amount of energy in the band 20–30 Hz as there is between 6 and 9 kHz, both having the same frequency ratio of 1.5.

Pink noise is frequently obtained directly from white by filtering, the filter reducing the white noise by 10 dB per decade as shown in the Figure.

CHAPTER 4. ROOM ACOUSTICS

We need only to remind ourselves of the difference between sounds heard in an enclosed swimming pool and those heard at home in the drawing or living room to realise that the environment has a profound effect. The difference between the two sounds apart from their relative loudnesses is due to *reverberation* (from Latin, to strike again), simply the multitude of reflections which add to the direct sound wave but slightly later in time. Under free-field conditions there are no reflections (Sect.1.4), therefore no reverberation, conversely within a room reverberation is present because reflections are inevitable. No reverberation at all is quiet and unnatural. Although this may therefore seem to be an impossible condition to achieve in practice (unless conversation were between two people suspended high up in the air), special *anechoic* (no echoes) rooms are constructed for experimental and calibration work, in these all sound is absorbed by the walls, floor and ceiling so none is reflected. Conversation or music within such a room gives rise to distinctly unfamiliar sounds. On the other hand too much reverberation (as in the swimming pool) is displeasing for in the extreme, reflected sound waves may still be around after the original has died away, such noticeable echo can even make conversation difficult.

Within this chapter the word "room" is used as a general term and although it may engender ideas of office or home, we include the larger spaces of lecture halls, cinemas, theatres, churches and cathedrals.

4.1 UNITS OF SOUND ABSORPTION

In Sect.1.4.1 we discovered how materials can be rated by a sound absorption coefficient. This allows comparison only so another unit for expressing actual absorption is also required. The unit is known as a *sabin* and is named after Professor W.C. Sabine (an American physicist) who was one of the earliest experimenters concerned with improving poor acoustic environments. The sabin is a measure of the absorption of a known surface area of a material which is completely sound absorbent (absorption coefficient = 1.0). Professor Sabine's work was published around 1895 so naturally the area was then quoted as one square foot, but now with metrication we change to an area of 0.0929 sq. metres (1 ft² = 0.0929 m²). Hence

Area of material (A)	Absorption coefficient (α)	Sound absorption (a)
1 ft²	1.0	1.0 sabin
0.0929 m²	1.0	1.0 sabin
0.0929 m²	0.6	0.6 sabin
1.0 m²	0.1	$\dfrac{1.0}{0.0929} \times 0.1 = 1.076$ sabins

leading to the general formula

$$a = \frac{A\alpha}{0.0929} \text{ sabins} \quad (A \text{ is in square metres})$$

(there was, of course, no awkward figure in the denominator when measurements were in square feet).

As an example which enables us to see more clearly the use of the unit, tests have shown that a person wearing a coat and sitting as a member of an audience has an absorption of about 5 sabins (averaged over the frequency range), meaning that if each of the surfaces (face, hair, coat, etc.) could be treated by the formula as above, then their sum would be a total sound absorption of 5 sabins.

Fig.1.9 shows how absorption coefficients can vary with frequency and we must therefore expect sound absorption to do likewise. As a practical example a carpet 3.097 x 3.0 m with an absorption coefficient, α = 0.1 at 120 Hz and 0.7 at 5000 Hz absorbs

$$\frac{3.097 \times 3}{0.0929} \times 0.1 = 10 \text{ sabins at 120 Hz},$$

and

$$\frac{3.097 \times 3}{0.0929} \times 0.7 = 70 \text{ sabins at 5000 Hz}.$$

Because the sabin by nature of its arrival long ago is associated with the square foot, it has no place as a S.I. unit nor even in metric calculations unless accompanied by the conversion factor. Hence we need not be too concerned with the unit but it is worth knowing of it in case it arises elsewhere. The unit which is developed from the idea of the sabin and which fits in with the metric system we will call the *square metre absorption unit* for lack of a shorter label and it is similar to the sabin in use except for being equal to the absorption of one square metre of completely absorbent material. We might even picture the effect as equivalent to an open window of 1 sq. m area in a room. All sound waves arriving over this area pass through the window and as far as the room is concerned, are lost, equivalent to total absorption (of course, assuming that no sound energy is reflected from outside back in). A square metre absorption unit is equivalent to

$$\frac{1}{0.0929} = 10.76 \text{ sabins } (1 \text{ m}^2 = 10.76 \text{ ft}^2).$$

4.2 REVERBERATION TIME

Consider a source of sound with a listener nearby within the same room. The first sound waves to reach his or her ears travel directly. These are followed by waves reflected by walls, floor, ceiling and objects but at lower levels because of absorption at each reflection (Sect.1.4). The time delay of each reflected wave depends on the total distance of travel from source to listener, in many cases several times that for the direct path. Thus the sound level experienced by the listener builds up and eventually reaches the *equilibrium level* when the energy is absorbed as fast as it is generated. Under anechoic conditions there is no energy build up from reflections which is why a talker at normal level appears to be speaking quietly. If now the sound suddenly ceases, sound waves in transit complete their journeys, those over the direct path doing so first with those over reflected paths arriving later, the level of each separate wave being determined by the absorptions it has experienced and the time of arrival by the total path length. The total sound thus dies away and the time for the intensity to fall to one millionth of its initial value on cessation of the sound source is called the *reverberation time*.

Here we are concerned with sound intensity, not pressure, hence for decibels we use 10 times the logarithm of the ratio and define reverberation time as that which is taken for the sound intensity to fall by 60 dB $\left(10 \log \dfrac{1,000,000}{1} \right)$ from the equilibrium level. On studying the various formulae it would become evident that theoretically the sound never completely dies away, it just continues to get less and less, however at 60 dB attenuation, for all practical purposes it has gone. As a reminder we use the abbreviation RT_{60}.

Before we get involved in graphs and formulae it is obvious that reverberation time must vary directly as the volume of a room for the reflected waves have farther to travel in a larger room and inversely as the sound absorption because they are attenuated more rapidly. The earlier example of the enclosed swimming pool compared with the home drawing room illustrates the point.

The growth of sound intensity to the eqiilibrium level and subsequent decay after cessation of the sound in a particular room is illustrated graphically in Fig.4.1(i). This is for an omnidirectional (omni is from Latin, all) sound source and the curve is theoretical and therefore considerably smoother than a practical one. The formulae for such calculations are rather complicated but we might have a feeling from experience with

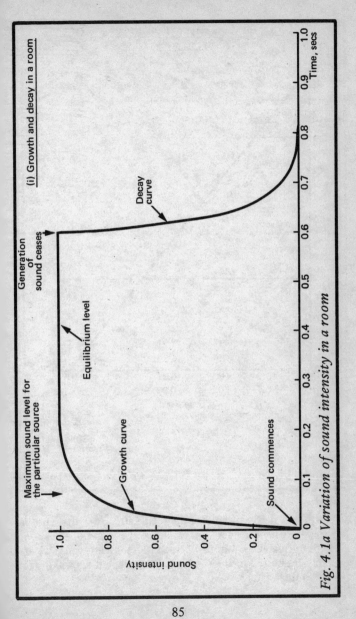

Fig. 4.1a Variation of sound intensity in a room

(i) Growth and decay in a room

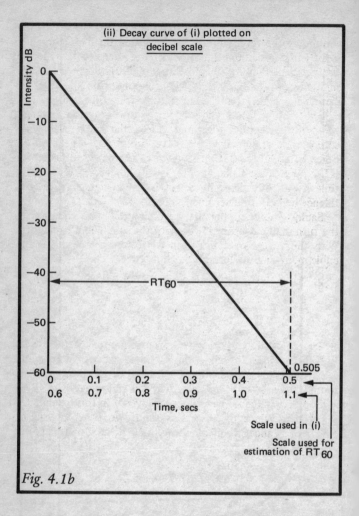

Fig. 4.1b

growth and decay of voltage on a capacitor[(2/2.2)] that epsilon creeps in, which in fact it does, hence the shape of the curve. The room details are not of great interest here because we are merely interested in the shape of the curve, but we will in fact refer to them later.

From the curve in Fig.4.1(i) there is no hope of even estimating the reverberation time because on such a graph it is difficult to read the time at which the intensity has fallen to 0.01, and certainly impossible at 0.000001 which is the one millionth or 60 dB point. By changing to decibels however we can do much better and again show how useful the dB system can be. By so doing and scaling the time from the moment when the generation of sound ceases, the graph becomes a straight line as shown in Fig.4.1(ii) and it reaches the 60 dB point at about 0.505 secs, which is therefore the value of RT_{60} for this particular room. This is not how reverberation time is normally determined but is shown here to give confidence in the formulae which follow.

Sabine published the first reverberation time equation at the turn of the century in terms of the square and cubic foot. With the advent of metrication we now use the square and cubic metre and the formula so adjusted is

$$RT_{60} = \frac{0.161\,V}{S\bar{\alpha}} \text{ secs}$$

where V is the volume of the room in cubic metres, S is the surface area of the boundaries in square metres and $\bar{\alpha}$ is the average absorption coefficient of all surfaces in the room (the bar over α indicates that it is an average).

Some inaccuracies were later discovered in this formula for the higher values of absorption and two other physicists, R.F. Norris and C.F. Eyring, modified it, now metricated as

$$RT_{60} = \frac{-0.07\,V}{S\log_{10}(1-\bar{\alpha})} \text{ secs}$$

(the minus sign looks odd but this is sorted out below).

We can put this to the test by calculating RT_{60} for the room of Fig.4.1. Its dimensions are: Volume $(V) = 85$ m^3, surface area [floor, walls and ceiling] $(S) = 120$ m^2, average absorption coefficient $(\bar{\alpha}) = 0.2$.

Then

$$RT_{60} = \frac{-0.07 \times 85}{120 \log_{10}(1 - 0.2)} = \frac{-0.04958}{\log_{10} 0.8} =$$

$$= \frac{-0.04958}{-0.09691} = 0.512 \text{ secs}$$

(the minus sign has disappeared).

This result agrees well with that obtained graphically from Fig.4.1(ii). While appearing simple enough, the value of $\bar{\alpha}$ is not always easy to determine because surfaces differ both in sound absorbency and in area. As an example, a room with wood-panelled side walls ($\alpha_s = 0.11$), curtained end walls ($\alpha_e = 0.45$), carpeted floor ($\alpha_f = 0.4$) and plaster ceiling ($\alpha_c = 0.03$) might be assessed as follows. Dimensions of the room are 6m x 4m x 2.5m high. Here the values of α are themselves averages over the audio frequency range but the calculation is more likely to be done at several different frequencies instead. Total sound absorption

$$a = \quad 2(6 \times 2.5)\,0.11 \qquad \text{(side walls)}$$

$$2(4 \times 2.5)\,0.45 \qquad \text{(end walls)}$$

$$(6 \times 4)\,0.4 \qquad \text{(floor)}$$

$$(6 \times 4)\,0.03 \qquad \text{(ceiling)}$$

$$= \quad 22.62 \text{ square metre absorption units}$$

$$\left(= \frac{22.62}{0.0929} = 243.5 \text{ sabins} \right)$$

but we stay with square metre absorption units because the formula is metric so the sabin is not appropriate. If absorption in sabins were known then we simply multiply them by 10.76 as mentioned in Sect.4.1

$$S = 2(6 \times 2.5) + 2(4 \times 2.5) + 2(6 \times 4) = 98 \text{ m}^2$$

$$\therefore \bar{\alpha} = \frac{a}{S} = \frac{22.62}{98} = \underline{0.231}.$$

Note the trap for the unwary here. $\bar{\alpha}$ is not simply $\dfrac{\alpha_s + \alpha_e + \alpha_f + \alpha_c}{4}$ as might be expected from the normal averaging of 4 numbers by dividing their sum by 4. In the acoustic case each α is associated with a different area and these must also be taken into account. The general formula is

$$\bar{\alpha} = \frac{\alpha_1 S_1 + \alpha_2 S_2 + \alpha_3 S_3 + \dots}{S_1 + S_2 + S_3 + \dots}$$

where α_1 is the absorption coefficient of area S_1, α_2 of S_2, etc.

Music itself takes many forms, generally needing a longer reverberation time than for speech to give a richness of tone. Most musical requirements lie in the range 1.0–2.0 seconds with dance bands at the low end of the range to symphony orchestras at the top end. Higher values up to 3.0 secs or slightly more are associated with church or cathedral organs. The types of room in which such sounds normally arise are commensurate with the appropriate reverberation time from the theatre with comparatively small volume and fairly absorbent surfaces (short time) to the spacious cathedral with stone walls and floors (longer time). For speech fairly short reverberation times are better, long ones bring confusion by returning one syllable to the talker during utterance of the next. A speech broadcasting or recording studio would be designed for a time of some 0.4 – 0.6 secs whereas school classrooms have a slightly higher value, say, 0.6 – 0.8 secs. Cinemas and theatres handling both speech and music are in the range 0.8 – 1.4 secs and in these cases the reverberation time is shorter with an audience than without because of the additional absorption.

4.2.1 Measurement

Several methods of measuring reverberation time have been developed, even the most elementary one of using a stop-watch but this has only any pretence to accuracy at the longer times. Modern techniques are most likely to employ a random (white)noise[3/2.2.2] source for the production of sound energy at all frequencies, feeding a power amplifier and loudspeaker. The sound is set up by this equipment and adjusted for level, then switched off. A measuring microphone which is followed by switchable filters to select the desired frequency band is connected to a timing device, usually with digital readout. This measures the time taken for a certain number of decibels decay. It is not necessary to wait for the full 60 dB for this may lead to difficulties with room noises, Fig.4.1(ii) reminds us that the decay expressed in decibels is directly proportional to time, therefore if, for example, the time for 15 dB decay is measured, then RT_{60} is simply 4 times this.

4.3 RESONANCES

Section 3.2.1 on vibration of air columns considers mainly what happens in tubes. The imagination need not be stretched much to realise that resonances will also arise in rooms even though the excitation is within the enclosure rather than at the end. Room air columns are effective within a space which is closed at both ends and arise between any two surfaces which can provide the necessary reflections, the different ways in which it happens being called *modes* (from Latin, modus, the way or manner). Of greatest significance therefore are the end wall—end wall (length) mode, side wall—side wall (width) mode and floor—ceiling (height) mode. A node exists at each wall giving a displacement pattern as shown in Fig.3.5(i) (this is for a string but the conclusion is the same) hence resonance occurs at the frequency for which the mode length equals $\lambda/2$. Thus is ℓ

is the length of the path or column, resonance arises at $\dfrac{c}{2\ell}$ Hz.

Considering, for example, a domestic room, say, 6 x 5 metres floor and 3 m to ceiling, the fundamental frequencies are: length mode, 28.7 Hz; width mode, 34.4 Hz; and height mode, 57.3 Hz, all of which are low. For larger rooms the resonances are even lower eventually becoming negligble compared with the effect of reverberation. Many other modes exist where a greater number of reflections occur, that is, not at 90° as is the case for the three direct ones mentioned above. Remembering that these are vibrating air columns and therefore have harmonics, some bass accentuation of sound at certain frequencies in small rooms is therefore inevitable although generally small. Nevertheless for the very best of high quality listening, special positions of loudspeakers to minimize the effect may be desirable. Such resonances are known as *eigentones* ("characteristic" tones from the German word "eigen").

An interesting conclusion is that room *shape* is important for minimum eigentone disturbance. Going to the extreme, a room which is cubical has length, width and height modes the same, all therefore giving rise to the same resonance, so tripling the effect. Hence we find that studies for example have special dimension ratios. In addition for minimum effect parallel surfaces are avoided so that sound waves only return to the point of origin after many reflections, so keeping both the eigentone frequency and its amplitude low.

CHAPTER 5.
ELECTROACOUSTIC TRANSDUCERS

"*Electroacoustic*" combines two words, "electricity" and "acoustics" so conjuring up in the mind electronic current in a wire on one hand and the sound wave on the other, very different entities indeed. Changing from one of these forms to the other requires a transducer (from Latin, to lead across), in audio known more generally as a microphone or loudspeaker depending on whether the move is acoustical to electrical or vice versa. These devices are an essential ingredient of audio engineering and are the links with the human being so warranting a chapter on their own even though some brief notes on most types are included earlier in the series.[5/1.7]

5.1 MICROPHONES

A microphone converts sound waves into electrical energy and as far as possible the output waveform should be a true replica of the input. Such energy transformation invariably involves in addition a mechanical conversion thus:

sound wave → mechanical energy → electrical energy

and as with electrical systems the coupling between each stage is important although generally it is unavoidably inefficient, for energy is being transferred between media of very different densities such as of air and metal. The mechanical conversion in the ear begins with the eardrum, this is a *membrane* (from the Latin, parchment), in most cases in the microphone it is a *diaphragm*, a very thin sheet of material which is caused to vibrate by the sound wave and itself drives the electrical unit. There are several different techniques for mechanical to electrical conversion and it is therefore appropriate for us to look at the various microphones available in terms of the conversion methods employed. In a broader sense however, microphones are classified by the two ways by which the diaphragm is made to respond, (i) the back of the diaphragm is sealed off from the front so that its response is proportional to the *pressure* of the sound wave and (ii) the wave may have access to both front and back of the diaphragm so that it is caught up in the motion of the particles and its response therefore corresponds to the particle *velocity*. Why velocity is important will become clear later but a reminder is that generated e.m.f. is proportional to conductor velocity in an alternator.[2/1.1.2] The principle is also known as *pressure gradient* operation and this is explained in more detail later for a particular type of microphone which employs the principle (Sect.5.1.4.2).

We will note whether the pressure or velocity (pressure gradient) principle applies on examining each type of microphone.

5.1.1 Characteristics and Measurements

As mentioned, microphones are one of the audio links with human beings and it is the latter who must be satisfied. Thus only listening tests can ultimately determine whether reproduction of speech or music with a particular microphone in the chain is satisfactory and undoubtedly a hi-fi expert can easily detect shortcomings in any system. Generally however, pure-tone tests are employed because of their repeatability and simplicity compared with listening tests. With the former the sensitivity/frequency characteristic tells much about the microphone and enables the user to see the effects of resonances and damping. However, unlike most frequency-sensitive electronic components on which straightforward electrical tests can be made to determine performance, the microphone, having both acoustic and electrical features, is more difficult to assess even when using pure-tones mainly because a known, repeatable sound field must be set up as the testing input. At the output there is no problem, a voltmeter being all that is required. Several other microphone qualities are also important, these and measurement of sensitivity are discussed in the following sections.

5.1.1.1 Sensitivity

We might define *sensitivity* as how readily something responds to a certain stimulus, for a microphone the stimulus arises from a sound wave, the response is the voltage appearing at its terminals. Generally the measurement is in terms of the open-circuit output e.m.f. rather than the p.d. on a given load because the p.d. can be calculated for any load once the open-circuit e.m.f. and microphone impedance at the particular frequency are known.[1/3.3] Noise voltages which are due to thermal agitation and are generated in the microphone impedance[3/3.2.2] cannot be eliminated (except by lowering the temperature) hence microphone sensitivity should be as high as possible in order to provide an adequate signal/noise ratio. Sensitivity can be expressed in several different ways, but usually by

$$S_m = 20 \log_{10} \frac{\text{open-circuit output e.m.f.}}{\text{free-field sound pressure at reference point}}$$

where S_m is the microphone sensitivity. "Free-field" needs explanation. Normally there are sound reflecting and absorbing materials present in any practical environment (Sect.1.4). These modify the sound wave and the effect is usually frequency dependent but naturally it is desirable that such distortions should not arise. Anechoic rooms (no echoes) in which all sound wave reaching floor, ceiling and walls are absorbed and used in the few places where the considerable expense can be justified. Slightly less efficient but effective is the anechoic chest or box with all internal surfaces padded for sound wave absorption. Such techniques are capable of providing relatively free-field conditions over much of the audio frequency range.

A simple measurement technique is illustrated in Fig. 5.1(i). The microphone under test is set up within the "free-field" environment facing a high quality testing loudspeaker. A variable-frequency oscillator feeds pure tones to the loudspeaker via an equalizer[5/4.2.2.3] for a constant level of sound pressure with frequency. In series is an attenuator,[5/3.2.3] set so that the sound field at the reference

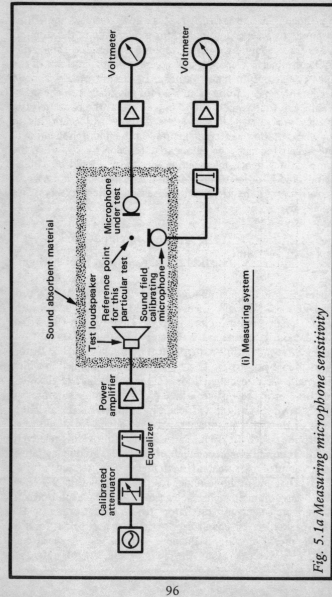

Fig. 5.1a Measuring microphone sensitivity

Sound absorbent material

Calibrated attenuator

Equalizer

Power amplifier

Test loudspeaker

Reference point for this particular test

Sound field calibrating microphone

Microphone under test

Voltmeter

Voltmeter

(i) Measuring system

96

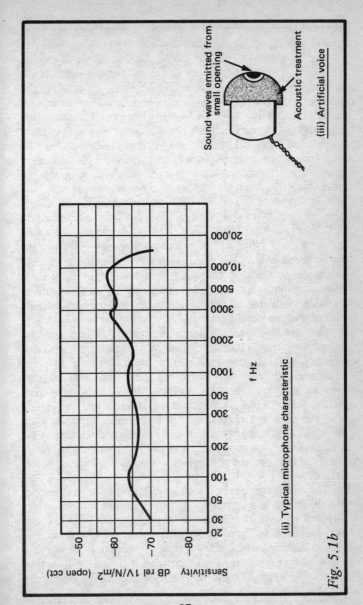

(ii) Typical microphone characteristic

(iii) Artificial voice

Sound waves emitted from small opening

Acoustic treatment

Fig. 5.1b

97

point is at the testing level of, say, 1 N/m² (10 dyn/cm²) and checked when necessary by a calibrating microphone system as shown. This also contains an equalizer to provide an overall flat response from the sound field pressure at the calibrating microphone to the voltmeter. The size and position of the calibrating microphone are arranged for minimum interference with free-field conditions. The microphone under test is connected to a high input impedance amplifier driving a voltmeter calibrated (usually in decibels) to read the e.m.f. By changing the oscillator frequency throughout the range required a graph of sensitivity/frequency as defined above is drawn, a typical one is shown in (ii) of Fig.5.1. As an example, at 3000 Hz the microphone sensitivity is −60 dB relative to 1 V/N/m². −60 dB rel 1 V = 1 mV, hence at a sound pressure of 1 N/m² at the reference point at 3000 Hz, the microphone output is 1 mV. Such a system as shown is quite adequate except that the work is tedious, thus fully auto- mated systems are available, consisting of a motor-driven variable oscillator at the input with moving paper chart recorder at the output.

For testing speech microphones such as are used in telephones the loudspeaker takes the form of an artificial voice, a flat response unit with an approximation to point source sound generation as for the mouth and with a shape and acoustic treatment to resemble the acoustic properties of the human head. A sketch is shown in (iii) of the Figure.

By expressing the sensitivity in this form, microphones can be checked for performance and different ones compared. Cumbersome and approximate as the test method may seem, there is at present no simpler method yielding similar information.

Thus all microphone sensitivity/frequency characteristics which follow are expressed in this way and the frequently adopted method of making 0 dB the level at 1000 Hz has been avoided because by so doing different types of microphone cannot be compared. A few words of warning however. It is essential to appreciate that all curves are *typical*, that is, they serve as an example only, no two microphones are exactly the same even though of the same model, moreover quite large variations result from different methods of construction. The characteristics therefore give us no more than a rough idea of the performance, nevertheless ample for our purpose.

5.1.1.2 Impedance

Impedance is easily measured using an impedance measuring set directly across the microphone output terminals, the set is quite likely to be based on the Wheatstone Bridge principle.[5/3.2.1] The impedance determines the type of amplifier input required. Low impedances range up to about 100 Ω, medium or line from about 100 to 1000 Ω and high above this up to about 100 kΩ, the highest being suitable for direct feed into FET stages.[3/2.5]

Low impedance microphones are matched to higher impedance amplifiers by use of a matching transformer.[5/3.2.2] There is additional cost and slight power loss by so doing but one great advantage of the low impedance unit is that longer microphone cables can be tolerated. Cable capacitance shunts the signal and although its effect is small across a low imedance it is capable of causing appreciable signal attenuation across a high one, moreover it is frequency dependent, giving rise to a greater loss at the higher frequencies. In addition noise picked up by a microphone cable develops smaller voltages across a low impedance.

5.1.1.3 Dynamic Range

Fig.1.7 indicates that speech read from a book with a microphone close (2.5 cm) gives rise to a sound pressure level of some 90 dB. With the microphone even nearer to the lips, levels of up to about 115 dB are experienced, for loud talking or singing, even higher. Orchestras too give rise to levels above 110 dB. Evidently a studio microphone must be capable of handling almost the entire range shown in the Figure, for other uses a smaller range may be sufficient, much depending on the particular type of use and likely distance of the microphone from the sound source. A microphone incapable of handling the louder sounds gives rise to amplitude distortion.

5.1.1.4 Directivity

Microphones are seldom equally sensitive to sounds arriving from all angles. A most convenient way of representing this directivity is by plotting a circular graph known as a *polar diagram* in which angles round the circle represent angles of sound direction from the microphone. The graph itself represents the microphone response at any particular angle by its distance from the centre. Fig.5.2 illustrates this, at (i) is the *directivity pattern* (also called the *polar response*) of an omnidirectional microphone. 0° indicates the direction in which the microphone is pointing. At (ii) is shown a bi-directional characteristic, here the response is equally good from directly in front and from directly behind but zero from the sides (the graticule of (i) is now omitted). The 90°−270° line is known as the null (amounting to nothing) plane. If the maximum response (at 0° and 180°) is R, then the response at any other angle is simply R cos θ as shown. More likely to be required is the unidirectional characteristic shown in (iii) for given this a microphone is able to discriminate against unwanted noise from the rear.

These are simply basic patterns, other shapes are obtained by special design or by use of two or more microphones together to combine their individual patterns.

5.1.2 Variable Resistance Microphones

The majority of these are also known as *loose-contact* or *carbon-granule* types because of their principle of operation. This is simply that of causing the sound wave to vary the resistance of many carbon contacts. The main features of such a microphone are shown in Fig.5.3(i). The case is not illustrated but it totally encloses the mechanism, thus the microphone is pressure operated. Several thousand tiny carbon granules (some 0.2 mm across, a special kind of anthracite is used) are poured through a hole in the bottom of a bowl-shaped chamber which is one of the two electrodes and is fixed to the frame. The hole is then sealed with a nylon plug. Dipping into the chamber is the moving electrode which itself is attached to a thin aluminium alloy diaphragm. A silk retaining washer ensures that the granules are confined

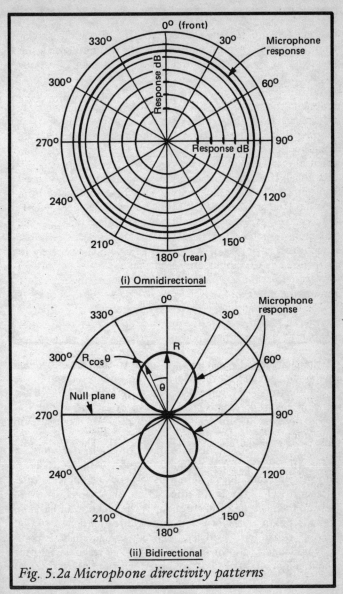

Fig. 5.2a Microphone directivity patterns

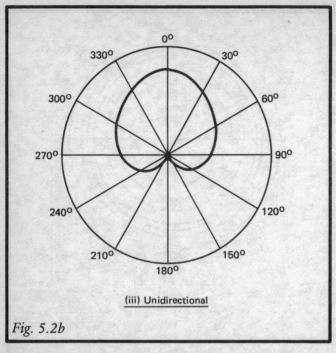

(iii) Unidirectional

Fig. 5.2b

to the chamber. Connexions are made to the frame to which
the diaphragm and moving electrode are connected and to the
fixed electrode. Movement of the diaphragm and hence the
moving electrode create pressure changes on the granules in
sympathy with the impinging sound wave. When pressure on
the granules increases the chamber resistance falls and vice
versa. To sense this resistance variation a direct current of
some 10 to 100 mA is required, connected theoretically as
shown but actually to terminals or sockets on the case.
Mathematical analysis of the action shows that a basic
feature in this type of microphone is the creation of harmonics, for this reason alone it is unusable for music. A second
reason is that the sensitivity/frequency characteristic is poor,
a typical one is shown in Fig.5.3(ii). The response falls off
above about 4000 Hz but compared with the other types of

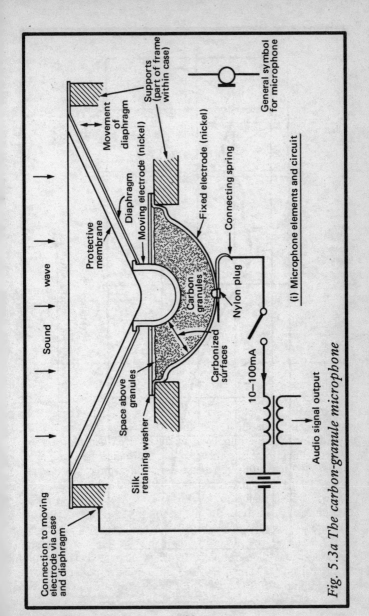

Fig. 5.3a The carbon-granule microphone

Connection to moving electrode via case and diaphragm

Sound wave

Movement of diaphragm

Supports (part of frame within case)

Protective membrane

Diaphragm

Moving electrode (nickel)

Fixed electrode (nickel)

Connecting spring

Carbon granules

Carbonized surfaces

Space above granules

Silk retaining washer

Nylon plug

General symbol for microphone

(i) Microphone elements and circuit

10—100mA

Audio signal output

103

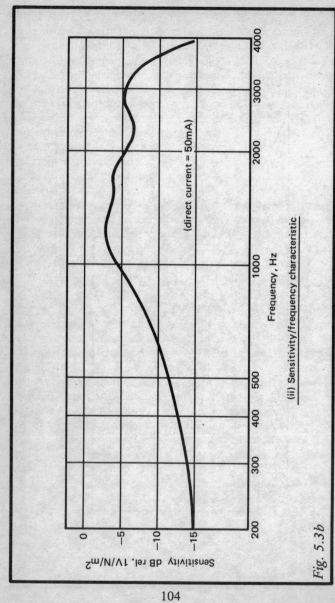

(ii) Sensitivity/frequency characteristic

Fig. 5.3b

104

microphone which follow, the overall sensitivity is high. It also varies directly as the direct current passing but there is a limit to the use which can be made of this because too high a current produces arcing between the granules with consequent deterioration of their surfaces. For telephone use which takes advantage of the high output and reasonable cost and for which the restricted frequency response can be tolerated, an average value of direct current might be taken as 50 mA.

The microphone impedance is almost purely resistive for it has no reactive components. Carbon has a negative temperature coefficient of resistance[1/3.4.3] hence as the granules warm up due to high direct current or prolonged use, the microphone resistance falls. The mean resistance also falls when the microphone is in use, for example by being spoken into in a telephone handset. Under the above conditions a modern microphone might have a resistance of 50–100 Ω, that is, in the low range. Because this type is used mainly for speech directly into the diaphragm, the directivity pattern is not of major consequence.

5.1.3 Piezo-Type Microphones

Piezo is derived from a Greek word meaning "to press". It is associated in microphone design with the technique of coupling the diaphragm directly to a solid material which responds to pressure by a shift of electrons within it, so creating a potential difference. The *piezoelectric* principle using natural or synthetic crystals is well established, the *piezo-junction* using a semiconductor has been developed more recently.

We must keep in mind that microphones which are based on the principles of diaphragm pressure on a crystal are not necessarily "pressure operated" as earlier defined, this description refers to the action of the sound wave, not to a mechanical feature within the device.

5.1.3.1 Piezoelectric

From the above derivation of "piezo", this type of microphone works on the principle of electricity from pressure. The first materials exhibiting the effect which were and still are used are found naturally. Each has a particular chemical composition and visually a transparent ice-like appearance, generally known as a *crystal* with a definite basic shape peculiar to that material. A commonly met example having a simple crystal shape is that of common salt, its crystal is a cube which is six-sided and regular. The crystal of alum also has a regular shape but 8 sides (octahedron). Common salt and alum do not have piezoelectric properties but Rochelle salt (sodium potassium tartrate – Rochelle is a French seaport) does and was one of the earliest naturally found materials used. Although all of its crystals are of the same shape, but not necessarily size, the shape itself is more complex than those for common salt and alum. Crystals may also be produced in the laboratory. For example, many chemical compounds are soluble in liquids, but only up to a certain degree, the solubility usually increasing with liquid temperature. When as much is dissolved as possible the solution is said to be *saturated*. If therefore a hot saturated solution of the compound is prepared and then slowly cooled, a small amount of the compound must be forced out of the solution and it appears in the crystalline form. If subsequently the liquid continues to evaporate, more of the compound is rejected and it is deposited on the existing crystals so that they grow continuously with their original crystalline shape maintained. Rochelle salt crystals may be produced synthetically in this way. The technique however is not suitable for all crystals used but is mentioned to show that we are not entirely restricted to mining natural materials. Some *ceramics* (pottery-like materials) also exhibit piezoelectric properties and several are now manufactured specifically for this purpose. Some of the better known piezoelectric materials are:

Sodium potassium tartrate (Rochelle salt) Quartz Ammonium dihydrogen phosphate (ADP) Lithium sulphate	crystal
Barium titanate Lead zirconate titanate	ceramic

Of these Rochelle salt is the most sensitive and cheapest but it has certain disadvantages. The salt is used dissolved in water as a medicine, so it is clear that humidity has an adverse effect on a crystal which accordingly must be well sealed against moisture. Little wonder therefore that it is being superseded gradually by ceramic elements with their much greater stability.

For ease of illustration we choose the ADP crystal, what is said about this particular one applies generally but with the more complex shaped ones it is often difficult to see the wood for trees. The ADP crystal is illustrated in Fig.5.4(i). It has three mutually perpendicular axes which we label as shown. The crystal is not used in its entirety but is sliced into small plates and the axis which is perpendicular to the main faces of the plate (hence parallel to the thickness) denotes the type of plate, that is, in Fig.5.4(ii) the plate is an *X-cut*. Identifying the axes of a plate is important so that the plate may ultimately be used with maximum sensitivity.

Considering a single X-cut plate as in Fig.5.5(i), if subjected to pressure at the centre as indicated, stress will be set up within the crystal tending to bend it along the axis AA'. Referring back to Fig.5.4(ii), with the stress along the Y axis, a voltage appears along the X axis, that is, on the major faces of the plate, for example, at any instant the top face might become +ve with the bottom −ve as shown. This is using a piezoelectric plate in the simplest mode. Many different arrangements are used most commonly perhaps in the *bimorph* system of two plates cemented together as illustrated in Fig.5.5(ii). For greater sensitivity larger stacks are used known as *multimorphs*. Each plate is sprayed on both faces with a metallic coating and in the bimorph the pair of plates is arranged so that when subjected to pressure the two inside coatings achieve the same polarity. The coatings are connected in parallel as shown, in the Figure the polarities are those for one direction of plate stress, reversing if the stress is reversed. Because of the parallel arrangement, the bimorph impedance is half that of a single plate, a desirable reduction as will be appreciated later. The bimorph may be used as a *bender*, usually by fixing the outside edges and applying

107

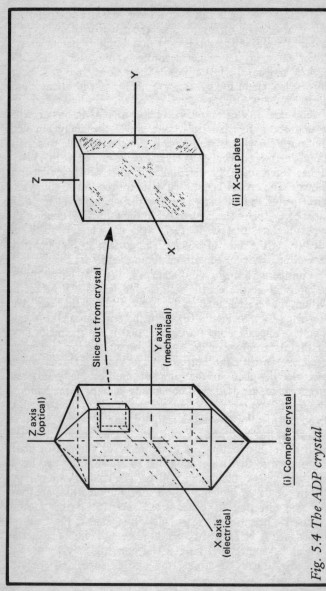

Z axis (optical)

Y axis (mechanical)

X axis (electrical)

(i) Complete crystal

Slice cut from crystal

Y

Z

X

(ii) X-cut plate

Fig. 5.4 The ADP crystal

108

pressure at the centre [Fig.5.5(i) and (iii)] or equally as a *twister* in which the unit is fixed at three corners with pressure applied to the fourth [Fig.5.5(iv)]. From the diagram the simplicity of construction is evident.

Notwithstanding this simplicity however, the response of these microphones can be made reasonably good. The output voltage is proportional to the degree of deformation of the crystal, thus there is little amplitude distortion. By careful design the amplitude of the movement of diaphragm plus plate (these are locked together) can be made fairly independent of frequency, this together with an output often at the millivolt level gives a useful general-purpose microphone with sensitivity/frequency characteristics typically as in Fig.5.6(i). The word "typically" must be emphasized because such microphones vary in price and quality over a wide range, the variation in characteristics is illustrated by the Figure. Generally, frequency response can be traded for level of output in that devices included to improve the response usually cause a drop in overall output. Comparison of Fig.5.6(i) with Fig.5.3(ii) demonstrates superiority of frequency response but lower level of output compared with that of the carbon-granule microphone.

Fig.5.5(ii) indicates the similarity in construction of a piezoelectric element with that of a capacitor,[1/4.3] in fact element capacitances vary from about 0.5 to a few nanofarads, the magnitude depending on the dimensions of a particular element and the crystal permittivity.[1/4.2] The impedance of the element is high, for example, at 1 nF it is 160,000 Ω at 1000 Hz. It is thus desirable that it should work into a high impedance load otherwise current drawn from the microphone creates in it an unwanted voltage drop. This is simply demonstrated by considering a microphone of internal impedance equivalent to a capacitance of 1 nF working through a 10 m length of low capacity microphone cable of 100 pF/m. As shown in Fig.5.6(ii) the output e.m.f. from the cable is half the generated e.m.f. of the microphone, a loss of 6 dB. 10 m of cable is perhaps a long length to consider but simplifies the explanation. Conditions are less onerous with a shorter length or with a change to coaxial

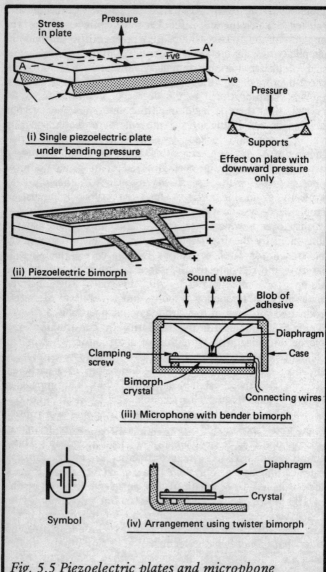

Fig. 5.5 Piezoelectric plates and microphone

cable[5/1.6.2] of lower capacitance, say, 50–60 pF/m. Clearly cable type and length are important with this particular type of microphone. However, an FET pre-amplifier can be built into the microphone case or an adjacent housing to avoid such problems, the input impedance to the FET circuit being very high[3/2.5] and the output impedance feeding into the cable lower.

The directivity pattern would be omnidirectional tending to be as in Fig.5.2(i) were it not for the microphone itself. The case interferes with sounds arriving from the rear (the *obstacle effect*, Sect.1.4.4), the attenuation it causes increasing as the wavelength falls and eventually becomes comparable with the case dimensions. Thus size and shape of the microphone, angle of incidence and frequency of the wave, all have an effect, the combined directivity pattern of Fig. 5.6(iii) might be taken as a very rough guide, that is, a fairly constant level of output at all frequencies to which the microphone is responsive at 0° but tailing off to the sides and to the rear as frequency increases.

5.1.3.2 Piezo-Junction

This is a microphone which could be mistaken for a piezo-electric type because of its similarity in both appearance and operation. It has however a transducer element functioning on a more recently developed principle. In place of the piezo-electric element is a semiconductor junction, for example, silicon planar[3/1.6.1] on which the diaphragm applies pressure to modify the transconductance. Semiconductor junctions are tiny and one of the problems in development is that there is little room for error in the point of application of the diaphragm pressure. Advantages to be gained are that the microphone is self-amplifying and is of fairly low impedance.

5.1.4 Electromagnetic Microphones

The first two of these described in this section are based on the fundamental principle of electromagnetism in which a conductor moving in a magnetic field has an e.m.f. induced in it.[1/5.3] In the microphone the conductor is either coupled to the diaphragm or is the diaphragm and its two

111

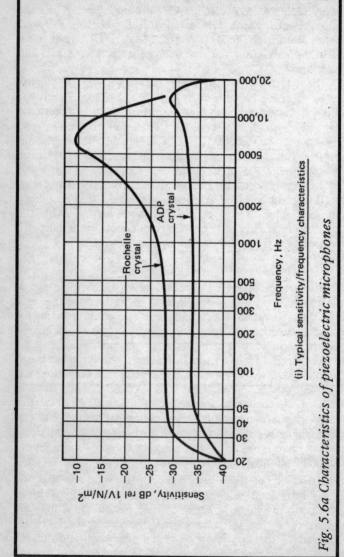

(i) Typical sensitivity/frequency characteristics

Fig. 5.6a Characteristics of piezoelectric microphones

112

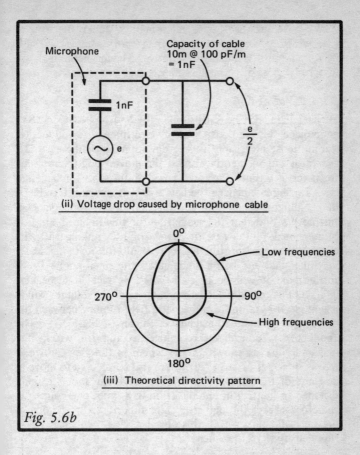

(ii) Voltage drop caused by microphone cable

(iii) Theoretical directivity pattern

Fig. 5.6b

ends provide the output. The two types considered are (i) the *moving-coil* which is pressure operated and (ii) the *ribbon* which is one of the few examples of velocity operation.

Following on from the above principle, a magnetic flux changing around a stationary conductor also generates an e.m.f. and this is the basic system of *variable reluctance* (also known as *moving-iron*) types. These are pressure operated and although in no way can they be classed as high quality microphones, they have various uses in speech systems.

We have grouped these microphones under the general title *electromagnetic* but the term *electrodynamic* will also be found and it is equally descriptive. The magnetic fields are invariably those of permanent magnets.

5.1.4.1 Moving Coil

Michael Faraday, the English chemist and physicist, in 1831 announced his famous law of electromagnetic induction leading most noticeably to the development of the electrical generator but incidently also to the moving coil microphone for it too is a generator and is based on the same fundamental law. In simple language the law says that an e.m.f. is induced in a conductor whenever it cuts magnetic flux. Going one step further we find that the magnitude of this induced e.m.f. is proportional to the *rate of change* of flux linking with the conductor and it is instructive to look into this mathematically in order that we better understand the microphone itself, furthermore it will be of help when we meet loudspeakers later. Let us consider a length (ℓ metres) of a conductor which is embraced by a uniform magnetic field running between the poles of a permanent magnet as shown in Fig.5.7(i). The conductor is considered to move to and fro within the magnetic flux as shown in (ii) where in the view shown it moves sideways or as in (iii) where it is considered to move in and out of the paper. Suppose the conductor moves a small distance (d metres) in one direction in a time t seconds. It therefore sweeps through an area of flux of $\ell \times d$ and denoting the total flux within this area by Φ, then since $\Phi =$ flux density (B) x area then

$$\Phi = B\ell d$$

and because the induced e.m.f. is proportional to the rate of change of flux linking with the conductor:

$$\text{e.m.f. induced} = \frac{B\ell d}{t}$$

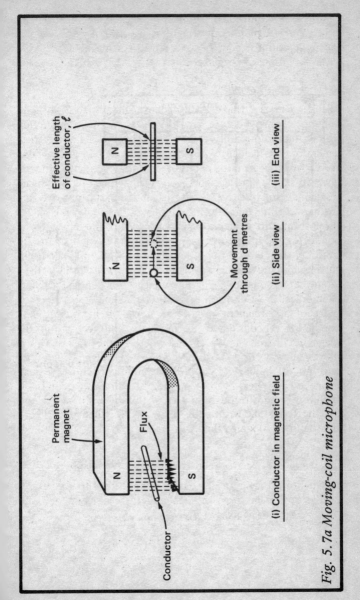

(i) Conductor in magnetic field

Permanent magnet

Flux

Conductor

(ii) Side view

N

S

Movement through d metres

(iii) End view

N

S

Effective length of conductor, ℓ

Fig. 5.7a Moving-coil microphone

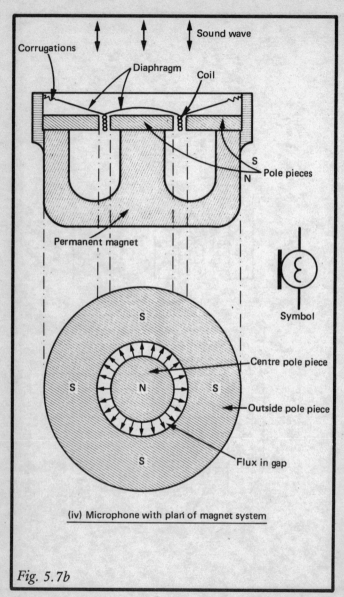

Sound wave

Corrugations

Diaphragm

Coil

Pole pieces

S
N

Permanent magnet

Symbol

S

S   N   S

S

Centre pole piece

Outside pole piece

Flux in gap

(iv) Microphone with plan of magnet system

Fig. 5.7b

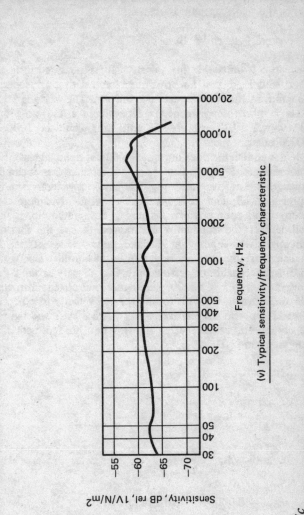

(v) Typical sensitivity/frequency characteristic

Fig. 5.7c

d/t is the distance moved by the conductor in the time t, that is, the velocity, call this v, then

e.m.f. induced $= B\ell v$

Now, in S.I units the flux density B is measured in teslas (shorthand T, $1\,T = 1\,Wb/m^2$)$^{(1/5.3)}$ and with t in seconds, the e.m.f. in the above equation is in volts. The working value of B in modern permanent magnet systems varies from about 0.1 to over 1.0 T (in pre-S.I. units this would be 1000 — 10,000 gauss).

If the conductor does not cut the flux at right angles but is at some angle θ relative to the flux direction, then it can be shown that e.m.f. induced $= B\ell v \sin\theta$ a general formula applying to all generators. We can conveniently ignore this latter complication however because in both microphones and loudspeakers the conductor is arranged to cut the flux at right angles whereupon $\theta = 90°$ and $\sin\theta = 1$.

Just to see the formula in action let us consider a permanent magnet capable of setting up a flux density around the conductor of 0.5 T (5000 gauss) with a 1.59 cm diameter coil of 20 turns. We wish to find the coil velocity to generate an e.m.f. of −66 dB relative to 1 V (−66 dBV − the figures are chosen for ease of calculation). Let e represent the generated e.m.f. Then

$$\text{No. of dB} = 20 \log_{10} \frac{1}{e}$$

$$\therefore \frac{66}{20} = \log_{10} \frac{1}{e}$$

$$\therefore \text{antilog } 3.3 = \frac{1}{e}$$

$$\therefore e = 0.0005\,V \quad (0.5\,mV).$$

Length of conductor, $\ell = 20 \times 1.59 \times \pi$ cms $= 1$ metre. Since $e = B\ell v$,

$$v = \frac{e}{B\ell} = \frac{0.0005}{0.5 \times 1} = 0.001 \text{ m/s} = 1 \text{ mm/s}$$

thus if the coil is moving at a velocity of 1 mm per second, an e.m.f. of −66 dBV is generated. From this we realise that for a uniform frequency response (that is, constant microphone e.m.f. for constant sound pressure at all relevant frequencies) the relationship between sound pressure and coil velocity must be constant, design for this is one of the important features in production of high quality units.

Next we see what the formula tells us about the basic moving coil microphone arrangement as shown in Fig.5.7(iv) which shows how when the diaphragm is caused to vibrate by an incoming sound wave the turns of the moving coil cut the flux in the gap between the pole-pieces of a permanent magnet. The circular magnet is one of high flux density with pole-pieces of high permeability so that the reluctance[1/5.2.2] of the magnetic circuit is low, hence creating the maximum flux in the annular (ring-like) gap. The radial arrows representing the flux in the plan of Fig.5.7(iv) show that a circular wire moving up and down in the gap cuts the flux at right angles and will therefore have maximum e.m.f. induced in it. The length of conductor ℓ in the formula is kept large by winding the coil of several turns, each turn can be considered as being in series with the next, therefore their e.m.f.'s are additive. The coil consists of round or flat very fine aluminium wire fixed to a thin (about 0.01−0.22 mm) aluminium alloy or plastic diaphragm with concentric corrugations at its periphery to avoid restriction of movement. The masses of the coil and diaphragm are therefore low, an essential requirement for good high frequency response.

Generally it is found that the coil + diaphragm system has a resonance at about 400 Hz or below, a second mechanical resonance also occurs between 4 and 6 kHz. In Sect.3.2.3 we saw how stretched membranes and circular plates have a frequency of resonance which is used to advantage for music. Microphone diaphragms have a similarity with these and also resonate but this is usually undesirable because the sensitivity is increased unduly around the resonance point. Sometimes however, advantage is taken of the effect to bolster-up a sagging response curve. The resonances are usually reduced by special cavities, air leaks or acoustic resistances of special fabric or felt within the case, in a way to provide an acoustic damping resistance so reducing Q (Sect.1.1.1) and therefore sensitivity peaks at the resonant frequencies.(2/3.7.1) With such treatment a reasonably flat sensitivity/frequency characteristic is obtained, typically as shown in Fig.5.7(v), generally somewhat lower in output level than given by a piezoelectric type [Fig.5.6(i)].

Consisting solely of several turns of fine wire, the coil impedance is low, say, 10–30 Ω and inductive. A step-up transformer to match to line amplifier inputs (say, 600 Ω) is frequently used and even built into the case although special amplifier input arrangements are possible. A particular problem affecting this type of microphone is its sensitivity to alternating current fields at 50 or 60 Hz which induce *hum* voltages in the moving coil. To reduce the effect, a *humbucking* coil is employed. This is a small coil, wound in the opposite direction and connected in series with the main one and located near it. Equal hum voltages induced in the two coils then cancel out.

Useful where a low impedance microphone is preferable, this type is generally more stable and robust than the piezoelectric there is no crystal to crack or as in the case the Rochelle type no likelihood of damage through moisture.

5.1.4.2 Ribbon

Although based on the same principle as the moving-coil, in construction this type is quite different. It has a single ribbon-shaped conductor instead of a coil, the shorter length therefore generating a lower e.m.f. By looking at the construction first we can then better appreciate some of the technical features. Fig.5.8(i) shows the basic ribbon microphone. The ribbon itself for a general-purpose studio unit might be of soft aluminium some 0.6–0.7 cm wide and 2.5–3.0 cm long, extremely thin, hence fragile. It is corrugated as shown to allow motion forward and backward and being very thin, responds well to transients, i.e. short duration

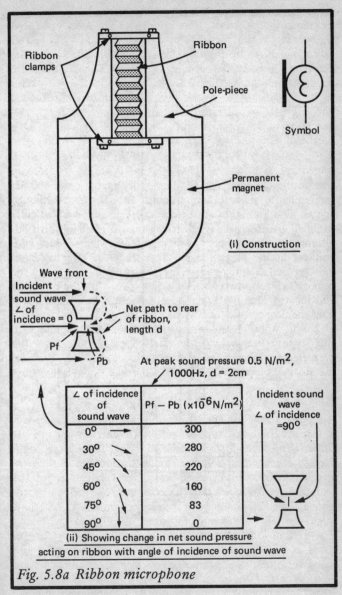

Ribbon clamps

Ribbon

Pole-piece

Permanent magnet

Symbol

(i) Construction

Wave front

Incident sound wave

∠ of incidence = 0

Net path to rear of ribbon, length d

Pf

Pb

At peak sound pressure 0.5 N/m², 1000Hz, d = 2cm

∠ of incidence of sound wave	Pf − Pb (×10⁻⁶N/m²)
0° →	300
30° ↘	280
45° ↘	220
60° ↘	160
75° ↓	83
90° ↓	0

Incident sound wave ∠ of incidence =90°

(ii) Showing change in net sound pressure acting on ribbon with angle of incidence of sound wave

Fig. 5.8a Ribbon microphone

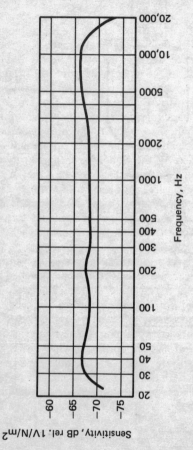

(iii) Typical sensitivity/frequency characteristic

Fig. 5.8b

122

signals with steep wavefronts. Clamps hold the ribbon within the gap between two pole-pieces extending from a permanent magnet of high flux density, say about 0.6 T or more. The gap accommodates the ribbon with less than 0.5 mm free space each side of it, this gives safe clearance for the ribbon while keeping the total gap as small as practicable to maintain the highest flux density ($e = B\ell v$, i.e. e is proportional to the flux density B).

Usually both sides of the ribbon are accessible to the sound wave but because the wave reaching the back of the ribbon has a greater distance to travel its effect on the ribbon is out of phase with that of the wave at the front, the magnitude of the phase difference depending on the lengths of the various paths existing between front and back. It is the difference between the front and back pressures which moves the ribbon and the net pressure multiplied by the ribbon area gives the total force acting at any particular time. It can be shown that the velocity attained by the ribbon is proportional to this force and hence to the particle velocity of the wave. We can look at this diagrammatically and also study more carefully the directivity pattern by use of Fig.5.8(ii), perhaps also giving us a useful exercise showing how mathematics can help in confirming some of the less tangible phenomena. The drawing shows at the left the ribbon and pole-pieces in plan with an incident sound wave generating pressures p_f and p_b at the front and back respectively, p_b having travelled farther over a net distance, d . The phase difference between p_f and p_b is a function of both d and the wave frequency. The actuating sound pressure ($p_f - p_b$) has been calculated for a 1000 Hz wave and d = 2 cms . The figures shown are very approximate and calculation is somewhat complicated, hence only the results are shown. The table shows how the net pressure varies with changing angle of incidence of the wave to the normal to the ribbon, running from maximum at 0° down to zero at 90° thus tending to confirm the directivity pattern for this type of microphone shown in Fig.5.2(ii). Discrimination against sounds at 90° is often an advantage in studio work.

The sensitivity is slightly below that of a moving-coil

microphone as would be expected from the shorter length of conductor. The frequency response however is good, extending from some 30 Hz to about 18 kHz as shown in Fig.5.8(iii). Sound waves operate the conductor directly and not via a diaphragm so giving low harmonic distortion. Transient response has already been described as good, in summary therefore the ribbon is a high-quality, low output device. Impedance is obviously very low, only a small fraction of one ohm, say 0.1−0.25 and mainly resistive. A step-up transformer is usually built into the case to increase both the output voltage and the impedance. Such a transformer is usually fitted with a magnetic shield to minimize hum pick-up. One major disadvantage is that the ribbon may be easily damaged, especially by strong air currents, in such cases therefore the microphone is best restricted to indoor work.

5.1.4.3 Variable Reluctance (Moving-Iron)

To be in tune with the title we might usefully recall what part *reluctance*[1/5.2.2] plays in magnetic circuits. A good analogy is with resistance in the electrical circuit where voltage has an equivalence with magnetomotive force (the driving force) and current with magnetic flux. Thus

$$\text{Electrical: } I = \frac{V}{R} \; ;$$

$$\text{Magnetic: } \Phi = \frac{F}{S}$$

or in words,

$$\text{Total flux} = \frac{\text{magnetomotive force}}{\text{total reluctance}} \; .$$

In its simplest form a variable reluctance microphone consists of a U-shaped permanent magnet with windings on the limbs and with a steel diaphragm held close to, but no touching, the two poles as shown in Fig.5.9(i), the pull of the magnet holds the diaphragm in position. Sound waves act upon the diaphragm, hence the term *moving-iron*. Tracing the magnetic circuit from, say, the magnet N-pole leads via air-gap g_1 , through the diaphragm, air-gap g_2 to the S-pole, thence through the magnet. Outside of the magnet the total reluctance consists of two reluctances in series and therefore additive (i) that of the diaphragm and (ii) that of the two air-gaps. Now the reluctance of the air-gaps is high compared with that of the metal path and like resistance it varies with length, $\left(S \propto \dfrac{\ell}{a} \right)$. $^{(1/5.2.2)}$ Thus when the diaphragm vibrates the total circuit reluctance varies accordingly because of changes in length of the gaps. Since $\Phi = \dfrac{F}{S}$, the total flux (Φ) varies and in doing so around the turns of the windings on the two magnet limbs, generates an e.m.f.

There are fewer restrictions on the number of turns in the windings compared with the moving-coil type, hence a large number can be used so producing a relatively high voltage and also impedance (several thousand ohms if required). The impedance in this case is highly inductive.

Many other arrangements are possible on the variable-reluctance principle. We can understand from Fig.5.9(i) that the diaphragm cross-section is a compromise between (i) being thick for maximum flux density and also to withstand the pull of the magnet and (ii) being thin for maximum sensitivity to sound wave pressure. Most better designs separate the two functions for increased efficiency and a *balanced-armature* design is shown in Fig.5.9(ii). Here a thin diaphragm moves a thicker armature. Most of the flux passes along the armature, varies as the armature moves and changes the gap reluctance thus inducing an e.m.f. in the winding.

Fig.5.9(ii) shows a typical sensitivity/frequency characteristic, a poor one compared with those of the microphones already studied and therefore of little value in studio work. Nevertheless the sensitivity is relatively high which makes this type particularly suitable for *sound-powered* telephones,

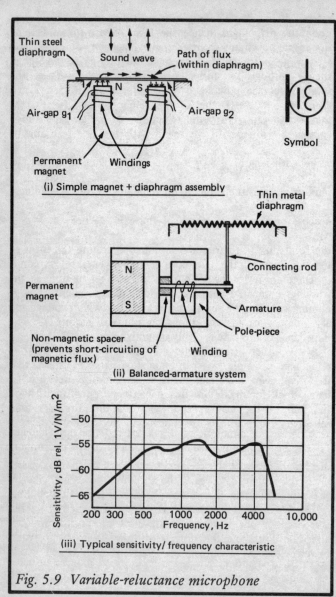

Fig. 5.9 *Variable-reluctance microphone*

for example, as used in army field work where a lusty voice overcomes the problem of having no power supply. It is also one of the several reversible transducers, that is, usable both as microphone and loudspeaker and it is therefore suitable for dictating machines and house telephones where a single transducer is switched for both talking and listening.

5.1.5 Electrostatic Microphones

Also commonly known as a *capacitor* microphone, this type is basically no more than a two-plate capacitor varying its capacitance by the effect of the sound wave on one of the plates so causing the thickness of the single dielectric to change. Recalling the relationship $C \propto \dfrac{A}{d}$ (1/4.2.3) where A is the plate area and d the separation between the plates (dielectric thickness), changes in d therefore produce corresponding changes in C. These are easily converted into similar variations in voltage. Electrostatic microphones are of studio quality. We look at the two main types: (i) the standard capacitor microphone (known as a condenser microphone in earlier days) which requires a d.c. or r.f. voltage supply for its operation and (ii) a modification of this which avoids the need of a supply, known generally as an *electret* microphone but still based on the capacitance principle.

5.1.5.1 *Capacitor*

It goes without saying that if one plate of a capacitor vibrates, the dielectric cannot be solid and the most obvious choice is air. Fig. 5.10(i) shows the elements of a capacitor microphone. The diaphragm which forms one plate is held very close to the back plate with the gap between them of some 0.025 mm. The diaphragm itself may take several forms, for example, a layer of evaporated gold on 0.05 mm thick glass, a 0.025 mm metallic foil (aluminium or gold) or metallized polyester film. In the latter case the dielectric is not wholly air since the film is also within the space between the two plates. The back plate is of solid metal such as nickel which may be drilled or slotted to improve air flow when the diaphragm moves. It is

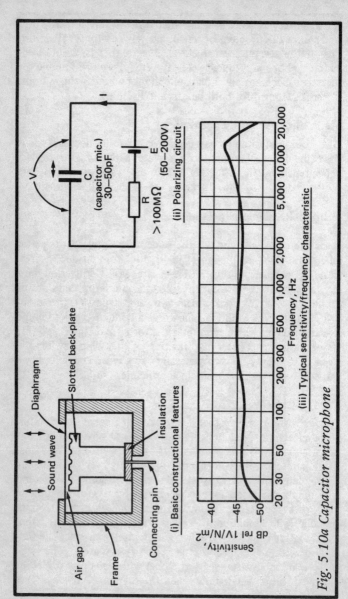

(i) Basic constructional features

Sound wave

Diaphragm

Slotted back-plate

Insulation

Connecting pin

Air gap

Frame

(ii) Polarizing circuit

C
(capacitor mic.)
30–50pF

V

I

R
>100MΩ

E
(50–200V)

(iii) Typical sensitivity/frequency characteristic

Sensitivity,
dB rel 1V/N/m²

–40
–45
–50

Frequency, Hz

20 30 50 100 200 300 500 1,000 2,000 5,000 10,000 20,000

Fig. 5.10a Capacitor microphone

128

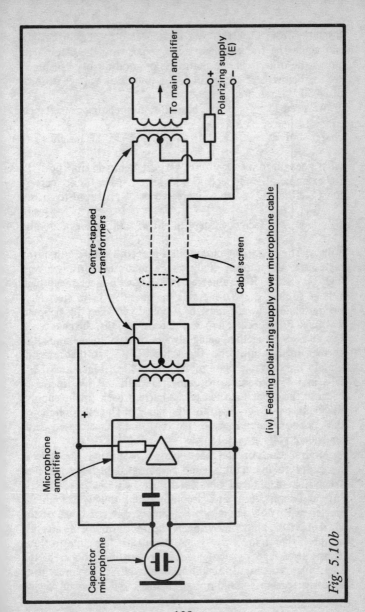

(iv) Feeding polarizing supply over microphone cable

To main amplifier

Polarizing supply
(E)

Centre-tapped transformers

Cable screen

Microphone amplifier

Capacitor microphone

Fig. 5.10b

insulated from the frame and is connected to one terminal or pin while the frame which is in contact with the diaphragm, forms the other. The insulation is of good quality because as we see below, a relatively high d.c. polarizing voltage of some 50–200 V is required.

We need to refer to the basic formula relating capacitance, charge and voltage, i.e. $C = \dfrac{Q}{V}$ (1/4.1) for an insight as to what goes on in a capacitor of varying dielectric thickness. If, for example, d is reduced, C is increased. Now for an increase in C either Q must be increased or V reduced to satisfy $C = \dfrac{Q}{V}$. In fact both happen but with the latter predominating. The electrical circuit in Fig.5.10(ii) shows that the plates are continuously charged, the circuit comprising a very high resistance, R connected in series with a relatively high voltage supply, E . To increase the quantity Q of electricity on the plates, current must be supplied yet when this happens a voltage drop occurs across R , hence V falls, the changes in Q and V together being always such as to maintain the correct relationship with the value of C at that instant. Should the movable plate of C shift so as to increase d , C falls and by the same reasoning Q falls but V increases because the capacitor then passes current back to the supply. By picking off V from the terminals of the microphone or the resistor, sound wave patterns are transformed into equivalent voltage variations.

Modification of the back plate to improve air flow is necessary to maintain a good response at high frequencies. When the diaphragm vibrates the air held as the dielectric suffers compression and decompression, thus developing a tendency to flow laterally. Put simply, the resistance of the air damps diaphragm movement, the opposition being greatest at the higher frequencies when the air has less time to flow. Thus the escape of air is assisted by cutting grooves in the back plate or by drilling holes through it, so preventing excessive pressure build-up. The slight reduction in capacitance is more than outweighed by the beneficial effect on high-frequency performance.

A high quality capacitor microphone has a good flat frequency response up to some 18—20 kHz as shown in Fig. 5.10(iii). Such a microphone may be quite small, generally tubular of 1—2 cm diameter, it does not therefore disturb the sound field appreciably.

Microphone impedance is high, almost purely capacitive, for example, at 1000 Hz a 40 pF unit has an impedance of about 4 MΩ. Capacity of even a short length of microphone cable would therefore swamp the microphone capacitance variations, thus it is general practice to install the pre-amplifier in the microphone case. The input impedance of the amplifier must be very high, for which an FET first stage is ideal,[3/2.5] powered usually from the microphone polarizing supply. The output of the FET amplifier can be arranged to be of medium impedance for feeding over a cable. A typical circuit using the phantom[5/4.1.2] of the cable to avoid separate wires is given in Fig. 5.10(iv).

The nuisance of having to supply a relatively high voltage over the microphone cable can be overcome by r.f. methods (i) by allowing the capacitance variations to modulate an r.f. carrier (at a few MHz) directly, then amplifying and demodulating[5/5.2] to gain the a.f. or (ii) by including the capacitor as part of the tuned circuit of an r.f. oscillator[3/3.3] (2 MHz or above), capacitance variations being converted to frequency changes, in fact producing a frequency modulated signal[5/5.3] which is subsequently amplified and passed to a discriminator for the a.f. signal to be extracted. Such systems, although seemingly complicated, can be housed entirely within the microphone case, even the power supply can be included which may be no more than a 1.5 or 3.0 V battery. The complete capacitor microphone then has terminals suitable for connexion to a normal microphone cable as with other types.

5.1.5.2 *Electret*

We see above that an essential requirement for an electrostatic microphone is that of a fairly constant charge, Q , on the plates of a variable capacitor. For the capacitor microphone the charge is rather inconveniently maintained by a relatively high voltage supply. A great improvement on this technique arises with the *electret* principle, simply that of implanting a permanent charge on the plates, thus being able to dispense with the voltage supply. One type of electret diaphragm is a metallized sheet of polycarbonate. This is heated and then given a permanent charge by surrounding it with a strong electrostatic field (for example, between other plates subjected to a very high voltage). The foil is cooled while still in the field and on removal the charge imposed on the foil remains.

In the microphone itself extremely high insulation between the diaphragm and back-plate must be provided otherwise the charge slowly leaks away. Some leakage is inevitable of course but with modern insulating materials this is so small that the microphone sensitivity should show no appreciable deterioration for many decades.

Usually the case contains an FET amplifier and miniature 1.5 V cell, the latter having a life of many thousand hours because of the low amplifier drain. The output level is as good as or better than that of the moving coil type. The frequency response however tends to be inferior in comparison with the capacitor microphone mainly because the diaphragm has to be slightly thicker and therefore less responsive to the highest audio frequencies. Nevertheless a fairly flat response from about 50 to nearly 18 kHz is obtainable from the better instruments. The built-in amplifier enables the designer to choose the output impedance, usually about 600 Ω.

5.1.6 Lavalier Microphones

Next a short digression into history. It concerns *lavalier* microphones, the small ones seen hanging from peoples' necks or clipped to the clothing. While such a microphone is simply a miniature version of one of those already described it may perhaps be of interest to see why they have such a French-sounding name and to look at just a few of the major features of design.

Long before our time a certain Duchess de la Vallière

132

adopted the custom of wearing a small jewelled locket or pendant on a chain around her neck. Because she was a very intimate friend of Louis XIV of France and therefore well regarded, this habit became fashionable and the hanging ornament was known as a *lavaliere*. Little did the Duchess know in the 17th Century that her name would be associated with microphones in the 20th.

The obvious advantage of a lavalier microphone is that a person may move about freely and that the distance between lips and microphone is always constant. Radio techniques wherein the transmitter and battery are also carried in the clothing and the signal is picked up "off-stage", add further freedom by dispensing with long microphone cables.

One of the design difficulties is caused by the fact that on talking, some sound emanates from the chest. While at a low level compared with that from the lips and therefore normally unnoticeable, when a microphone is actually placed on the chest this source of sound becomes of significance. Cavities in the human chest resonate at low frequencies and emphasize those particularly around 700 Hz, hence conducting more sound directly to the microphone. Intervening clothing may alter things somewhat and in addition clothing around the microphone will tend to reduce the higher frequencies (say, between 3 and 10 kHz) by absorption. A second dissimilarity with normal microphone operation arises from the fact that the microphone is in an unnatural position. Ears and microphones are normally directly in front of a talker, or singer, not below the lips with the chin intervening.

Added together, such abnormalities result in an unnatural quality. The cure perhaps is obvious, either design a microphone with lower response at about 700 Hz and rising response above 2–3 kHz or incorporate these features in an equalizer[5/4.2.2.3] which is connected in tandem with a normal microphone. The equalizer solution is preferable because it is impossible to design a lavalier microphone to suit all the different shapes of people and their clothing. For high-quality work a variable equalizer (a network which introduces attenuation or amplification at chosen frequencies) may be used to adjust for the best tonal quality, perhaps by direct

133

comparison with a normal microphone channel.

Electret microphones are especially suitable for lavalier working because of their small size, electromagnetic types are unsuitable because by their very nature they include relatively heavy and bulky permanent magnets.

5.2 LOUDSPEAKERS

In operation a loudspeaker is the converse of the microphone, it receives electrical energy and converts it into sound waves. In the process mechanical conversion is again employed thus the complete transformation involves:

electrical energy → mechanical energy → sound waves

A loudspeaker unit requires some kind of mounding, a *baffle* or a cabinet. These are not for aesthetic reasons only, their design is important in the final quality of sound output. Under the general term "loudspeaker enclosures" they are considered later in Sect.5.2.2.4.

Some microphone principles are reversible in that the particular arrangement for conversion of sound to electrical energy can also be employed to convert electrical into sound energy. For example, the moving coil microphone has its counterpart in the moving coil loudspeaker while the piezoelectric principle is also used in both types (but not the piezo-junction). To avoid duplication, references are made back to the appropriate sections on microphones as necessary.

5.2.1 Characteristics and Measurements

From Chapter 4 it follows that the sound heard from a loudspeaker in a room is modified by the acoustic qualities of the room itself. This is one reason why we must always be careful in interpreting loudspeaker characteristics because these are most likely to have been made under anechoic or near-anechoic conditions. Another reason for accepting characteristics as a guide only is that, as with microphones, no two loudspeakers are exactly the same, even of a particular manufacturer's type. But the ear is tolerant and two or three decibels change, even though seemingly a lot on paper, is hardly discernible, thus for the non-expert listener noticeable differences on graph paper between two loudspeakers may be undetected when comparing them in action.

Anechoic conditions as we have seen earlier are obtained by using specially constructed rooms. However, it is possible to make tests high in the air away from reflections (weather and extraneous noise permitting), it is the lowest frequency which determines the height of the loudspeaker supports, generally at least half a wavelength. For testing down for example to 30 Hz, the loudspeaker should be about 6 m high and be clear of all objects except the supports and testing microphone which cannot be avoided.

Designers and manufacturers make all kinds of complicated measurements, we consider a few of the most important:

5.2.1.1 Response

This is perhaps the most useful, the response in acoustic output for a given stimulus in electrical power, on a frequency basis. The loudspeaker under test is mounted in anechoic conditions at a certain distance, say, 1–2 m from a small calibrated microphone. By "calibrated" is meant that its sensitivity is known at all frequencies. It is however part of a complete calibrated microphone system as shown within the dotted lines in Fig.5.11(i) which illustrates the measurement arrangements. The equalizer is designed for the particular microphone so that the overall gain of the measuring system from sound pressure at the microphone to reading on the decibelmeter is constant over the frequency range concerned. A sinewave generator excites the loudspeaker via a power amplifier matched to it. A cathode ray oscilloscope monitors the system output to check purity of waveform. The input power to the loudspeaker must be constant at all test frequencies and given that the test loudspeaker impedance is known, this can be measured by the voltmeter shown. Unless overload tests are being made the input power must be within the operating range for the particular unit. More sophisticated equipment may be used, for example, a motor-driven sweep-frequency oscillator mechanically coupled to

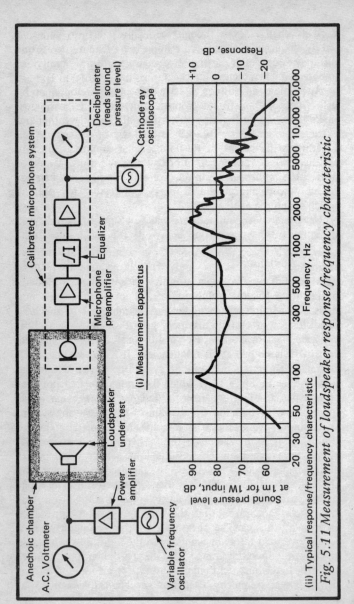

(i) Measurement apparatus

(ii) Typical response/frequency characteristic

Fig. 5.11 Measurement of loudspeaker response/frequency characteristic

136

drive a chart (moving paper) recorder. The response/
frequency characteristic can be expressed as sound pressure
output from the loudspeaker at a given distance or simply as
decibels change from some given frequency (e.g. 1000 Hz) as
shown typically in Fig.5.11(ii). When directional characteristics are required it may be more convenient to rotate the
loudspeaker rather than to move the microphone round it.

5.2.1.2 Impedance

We might get the impression that, say, a 4 Ω loudspeaker has
an impedance of 4 Ω irrespective of frequency. This is
certainly not the case, most loudspeakers have a pronounced
resonance at low frequencies and a rising impedance at the
higher ones. The impedance/frequency characteristic of a
loudspeaker is important because the impedance controls the
power which is accepted from a power amplifier and hence
affects the sound power output. Impedance/frequency
characteristics are discussed for the various types of loudspeaker in the following sections.

Measurement is made with the loudspeaker in action by
maintaining a constant current at all frequencies through the
unit and measuring the voltage across the unit terminals.

5.2.1.3 Efficiency

Not of great importance to most users because usually power
output from an amplifier is not at a premium but we should
have some idea of what the range is. As shown above the
sound pressure or s.p.l. is measured when response/frequency
characteristics are .produced, a relatively straightforward
process because the voltage output of most microphones
depends on the sound wave pressure actuating the diaphragm.
Sound intensity is not so easily measured but by referring
back to Sect.1.3.2 we see that calculation of it can be carried
out directly once the sound pressure is known so next follows
a practical example to illustrate the whole process. This will
also afford some more practice in handling acoustic quantities
and decibels. Some assumptions and simplifications are made,
this is in the very nature of acoustic analysis.

Consider a loudspeaker rated at 20 watts maximum

electrical input set up as in Fig.5.11(i) with the calibrated microphone 1 metre in front of it. The loudspeaker is mounted in its cabinet which takes the form of an infinite baffle (Sect.5.2.2.4 will show that all radiation is to the front). When supplied with 20 watts power at a certain frequency the sound pressure level is measured as 109.8 dB. We wish to calculate the efficiency of the loudspeaker as a transducer of electrical to acoustic power.

Fig.1.7 reminds us that, for any given sound wave, sound pressure and sound intensity *levels* have the same value, thus if p represents the sound intensity:

Sound intensity level $= 109.8$ dB

$$= 10 \log \frac{p}{10^{-12}} \text{ watts per square metre}$$

(refer to Sect.1.3.2 if in difficulty).

$$\therefore \frac{109.8}{10} = \log \frac{p}{10^{-12}} \quad \therefore \text{antilog } 10.98 = \frac{p}{10^{-12}}$$

$$\therefore p = (\text{antilog } 10.98) \times 10^{-12} = 0.0955 \text{ W/m}^2 .$$

This is the acoustic power flow at 1 m in front of the loudspeaker. The unit is also radiating in all other forward directions so to find the total acoustic power flowing at 1 m away we sum up for the hemisphere at 1 m radius. The surface area of a sphere is $4\pi r^2$, therefore for a hemisphere, $2\pi r^2$ which, when r = 1 m becomes 2π m^2 .

$\therefore$ Total acoustic power output $= 0.0955$ W/m^2 flowing through an area 2π m^2, i..e. $0.0955 \times 2\pi = 0.6$ watts.

$\therefore$ Loudspeaker efficiency $= 0.6/20 \times 100\% = 3\%$.

We know that motor car engines are inefficient, loudspeakers are many times worse! This is not a particularly poor loudspeaker either, very cheap ones are around 1% while good domestic loudspeakers reach some 5%.

The calculation is made over a hemisphere of 1 m radius only because the original s.p.l. measurement was made at this distance. It is logical to expect the same answer whatever the distance from the loudspeaker. We can put the inverse square law (App. A1.3) into service for confirmation, recalling that if the distance is doubled, the power flow is divided by 4, that is, reduced by approximately 6 dB (3 dB is half the power, 6 dB is one quarter). The sound intensity level at 2 m is therefore $109.8 - 6 = 103.8$ dB so that $p_2 =$ antilog $10.38 \times 10^{-12} = 0.02399$ W/m^2 . The larger area $= 2\pi \times 2^2 = 8\pi$ m^2 and total power $= 0.02399 \times 8\pi = 0.6$ watts — as before. A little further thought will show that this holds good whatever the distance at which the measurement is taken.

5.2.2 Moving Coil Loudspeakers

If we talk in terms of numbers then the moving coil type has pride of place in the loudspeaker world today, especially for domestic use. In construction there is a host of variations but the basic drive principle never changes. In Sect. 5.1.4.1 on moving coil microphones we looked at the general formula applying to all *generators* because in the microphone a sound wave generates electricity. Changing now to loudspeakers we need the general formula for *motors* for in these electricity creates a force for moving a cone or diaphragm. The force,[1/A9.2] F on a current-carrying conductor in a magnetic field is calculated from

$$F = B\ell I \sin \theta \text{ newtons}$$

where B is the flux density[1/5.2.4] in teslas (Wb/m^2),
ℓ is the length of the conductor in metres, and
I is the current in amperes.
θ is the angle between the conductor and the field.
As with the microphone, loudspeaker design aims at making θ as near as possible to 90°, therefore $\sin \theta = 1$. To see this formula in action we might consider what force would arise on the microphone diaphragm in the example given in the earlier section with a current of 100 mA flowing. The flux density is 0.5 T, conductor length 1 metre.

$$\therefore \ F \ = \ B\ell I \ = \ 0.5 \times 1 \times 0.1 \ = \ 0.05 \ N$$

meaning that the coil would apply a force of 0.05 newtons to the diaphragm tending to move it towards or away from the magnet system depending on the direction of the current. To get 0.05 N into perspective by recalling that the force of gravity on a mass of 1 kg is 9.81 N,[1/A9.2] it must be equivalent to a weight of about 5 gm (a little less than 1/5 oz). Loudspeaker coil currents are generally greater than 100 mA of course and an example of this is given later. Fig.5.12 illustrates how the basic principle is brought to fruition in a moving coil loudspeaker unit. As mentioned earlier, there are countless designs, this is one example only.

Bolted to the frame is the permanent magnet assembly, the particular model shown having a circular or *ring* ceramic magnetic. Many alternative magnet alloys are available comprising aluminium, cobalt, nickel or iron, the important factor being that the flux density across the gap in which the coil is situated should be high ($F \propto B$). Gap flux densities range from about 0.5 to 2.0 T. The pole-pieces which extend the magnetic circuit are of low reluctance[1/5.2] material such as a nickel-iron alloy. The cone which is of thin metal, special paper or plastic, sometimes with a fibre-glass or cloth base and usually circular but sometimes elliptical is attached to the frame at its outer periphery and to the coil former at the apex. The one shown is corrugated to allow forward and backward movement and apart from this must be rigid yet light to reduce the mass which the coil has to move. A centring device is essential to ensure that the coil former and coil are correctly positioned in the gap to avoid rubbing on the pole-pieces. The device is sometimes known as a *spider*, a corrugated one is shown in use in the Figure but many other designs are available, one particular alternative design is also shown. The coil itself is multi-layer of aluminium or copper wire or ribbon. We can now calculate the force arising in a typical practical unit in which the moving coil has a diameter of 3 cm and is wound with 40 turns of 0.112 mm diameter copper wire (1.75 Ω/m). The magnet system has a flux density of 0.9 T. Under test a dc power of 20 W is supplied. We wish

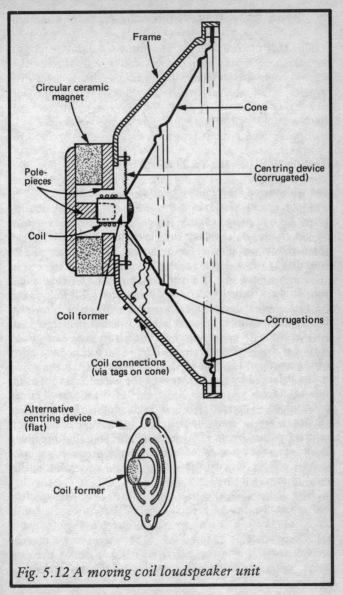

Fig. 5.12 A moving coil loudspeaker unit

141

to calculate the approximate force exerted on the diaphragm.

Length of wire in coil $= 40 \times \pi \times 3$ cm $= 3.77$ m

$\therefore$ Resistance of coil $= 3.77 \times 1.75 = 6.6\ \Omega$

Since $P = I^2R$, $I = \sqrt{\dfrac{P}{R}} = \sqrt{\dfrac{20}{6.6}} = 1.74$ A

Then $F = B\ell I = 0.9 \times 3.77 \times 1.74 = 5.9$ N,

and with such a force quite a large cone can be moved.

In design an allowance is made for the fact that flux and turns are not always exactly at right angles, that flux density in the gap is not constant and that there is fringing,[1/5.2.3] these effects are allowed for by adding a constant, k, to the formula (i.e. $F = B\ell Ik$), where k is less than 1.

A moving coil unit is a mechanical vibrating system and as such it has a natural resonance. From Sect.1.1 it follows that because small speaker cones have high stiffness and low mass the cone resonant frequency is high whereas the larger loudspeaker with its more floppy or compliant cone and greater mass has a low resonant frequency. At resonance the effect of the inertia of the mass (it takes power to get things moving) is just balanced by the liveliness of the compliance just as the two reactances X_L and X_C [2/3.1] of the electrical circuit balance out. Therefore with no mechanical resistance present vibration would continue indefinitely, but of course, such resistance does exist. This leads to the fact that because a loudspeaker has greatest efficiency at resonance, larger cone units are more suited to low frequencies, leaving the smaller units to handle the higher ones.

Now if the cone excursions are greater at or near mechanical resonance, the back e.m.f. due to motion of the coil in the magnetic field is correspondingly greater, therefore so is the impedance. Accordingly the impedance/frequency characteristic is anything but flat. Fig.5.13 shows the basic shape to expect for a low-frequency unit. It has a pronounced

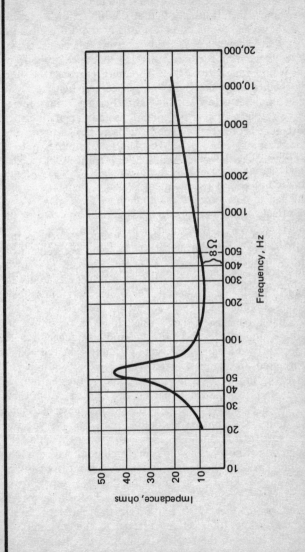

Fig. 5.13 Typical moving-coil unit impedance/frequency characteristic (8Ω)

peak at the resonance point and a steady rise thereafter as frequency increases because the coil is inductive. The nominal or *rated* impedance is usually that at the dip following resonance as shown and for an l.f. unit is usually at 400 Hz. Because of the wide impedance variation, correct matching to an amplifier is far from accurate but this is not such a disadvantage as would appear from the Figure. At resonance, because of the high loudspeaker unit impedance the signal current from an amplifier is low but the unit efficiency is high therefore to a certain extent these conditions counteract. Further improvement is obtained by good enclosure design to provide acoustic damping (Sect.5.2.2.4). The low amplifier output impedance is also an asset because it is effectively in parallel with the moving coil and therefore lowers the Q of the resonant system, so flattening the response.

5.2.2.1 Systems for Extended Response

Because the cone moves more slowly at low frequencies the acoustic power generated, which depends on the volume of air moved, is less. Compensation for this is by use of a larger cone. However, the mass of such a cone reduces its ability to respond to the higher frequencies for which therefore a small light cone is preferable. These conflicting requirements bring the inevitable conclusion that a single loudspeaker unit as in Fig.5.12 is unlikely to cover the full audio frequency range. Thus for hi-fi working, several types of loudspeaker combinations are used, the simplest to understand but not so simple in design is the use of a small cone attached to the centre of a large one, in effect the two in parallel and excited by the same drive. More commonly found is the double or treble loudspeaker unit system, taking the double as an example, consisting of a bass (l.f.) unit with large cone (up to 45 cm diameter) and sometimes known as a *woofer* in conjunction with a treble (h.f.) unit having a small, light cone and called a *tweeter*. A *crossover* network (or filter) preceding the two units directs the appropriate range of frequencies to each, for example, if the *crossover frequency*, f_c , is 1000 Hz, then theoretically all frequencies below 1000 Hz are directed to the bass unit, all frequencies above to the treble. In practice the changeover is in no way as precise as this.

5.2.2.2 *Crossover Networks*

We start with the simplest design, quite usable in practice, although unsuitable for truly hi-fi work. Nevertheless given loudspeaker units of moderately wide frequency response, good results may be achieved. The network consists solely of an inductor to feed power to the l.f. unit in parallel with a capacitor feeding the h.f. unit. The inductor passes low frequencies while rejecting the high ones whereas the capacitor passes high frequencies but not the low.

Crossover networks are subdivided into three classes according to their rate of change of attenuation with frequency in the crossover region and hence according to their configuration. They are known as *first*, *second* or *third order* networks and have 6, 12 and 18 dB attenuation per octave respectively (these terms are illustrated in Fig.5.14). The one we have in mind is of the first order and the practical network is shown in Fig.5.14(i), L and C being the inductor and capacitor mentioned above. The two loudspeaker units have similar nominal impedances and this value is taken as the input impedance of the network. The simple formulae used avoid impedance phase angles and accordingly the unit and the network input impedance are considered as a single design resistor R_0 .[5/3.3] Then:

$$L = \frac{R_0}{2\pi f_c} = \frac{0.159\, R_0}{f_c}$$

$$C = \frac{1}{2\pi f_c R_0} = \frac{0.159}{f_c R_0}$$

so having chosen f_c according to the response characteristics of the two units, determination of L and C is straightforward.

The attenuation/frequency slope of second-order networks is shown in Fig.5.14(ii) which when compared with (i) shows the superiority in rate of cut-off. The network itself may have

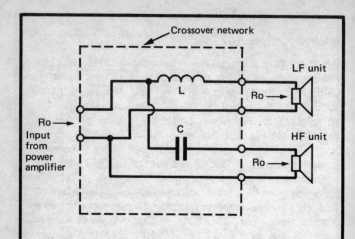

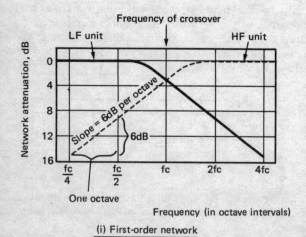

(i) First-order network

Fig. 5.14a Crossover networks

146

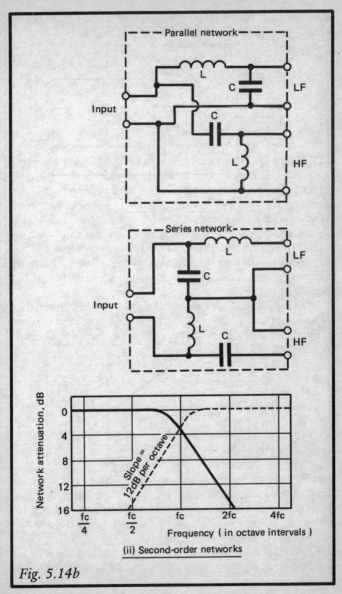

(ii) Second-order networks

Fig. 5.14b

either of two forms, with the units in series or parallel as shown. The design formulae are:

$$L = \frac{R_0}{\sqrt{2}\pi f_c} = \frac{0.225\,R_0}{f_c}$$

$$C = \frac{1}{2\sqrt{2}\pi f_c R_0} = \frac{0.1125}{f_c R_0} \ .$$

This is a network in which both inductors have the same value and also the capacitors but in modifications of the type they may have different values. The third-order network has additional components and even more complex networks in the form of complete section filters.[5/3.3] Alternatively active circuits using operational amplifiers[3/4.3.1] are employed, so much depends on the quality of the response required and the characteristics of the loudspeaker units. Such additional sharpness of cut-off however is only necessary for highly specialized work.

Working at audio frequencies brings the expectation that component values may be relatively large for example, for 8 Ω units with a crossover network as in Fig.5.14(ii) and crossover frequency 2 kHz

$$L = \frac{0.225 \times 8}{2000}\ H = 0.9\ mH$$

$$C = \frac{0.1125}{2000 \times 8} = 7.03\ \mu F\ .$$

Because the resistance of the inductor must be kept low (the formula does not even allow for it), a fine gauge winding cannot be used and because no direct polarizing voltages are present as in transistor circuits, C cannot be an electrolytic, thus both components tend to be bulky.

Efforts to achieve a flat loudspeaker system response do not stop at a single crossover network, two may be employed with three loudspeaker units to cover the range, one each for low, middle and high frequencies. With such systems practically the whole audio range can be covered with no more than a few decibels variation.

The whole idea is summed up pictorially in Fig.5.15. The response characteristics in (i) and (ii) are to the same scale and are those of the two loudspeaker units to be used. They both have the same maximum response which is maintained at 3000 Hz. This frequency is therefore chosen for f_c. The attenuation/frequency characteristic of a suitable crossover network with 12 dB/octave slope is shown in (iii). To determine the overall characteristic of (iv) we have therefore to add the responses of the two units on a sound power basis (we must not fall into the trap of adding decibels in this particular case) and from this net response, reconverted to decibels, subtract the crossover network attenuation — all this at each frequency needed for plotting the graph. The result in (iv) is that of a system covering the greater part of the audio range, a very good response indeed. It is also indicative of the fact that a 3-unit system would almost bring perfection as far as the response/frequency characteristic is concerned.

5.2.2.3 Horns

These are certainly not devices recently invented for use with loudspeakers, they were introduced long ago for improving the sound output of the voice (the megaphone) and of some musical instruments, also in the opposite direction for increasing the sound input to the ear in cases of deafness. Technically a horn works by improving the matching between two very different acoustic impedances, it is in fact an *acoustic transformer*. In a loudspeaker we can look upon the open air as being free and easy to move, therefore of low acoustical impedance compared with the diaphragm which is of dense material and much more tightly held, less movable and therefore of high impedance. The horn is of small diameter at one end, called the *throat*, tapered to a large diameter at the other end, the *mouth*, names consistent with the acoustical transformation in the human vocal tract. In loudspeakers the horn is driven at the throat by a cone of similar diameter, the

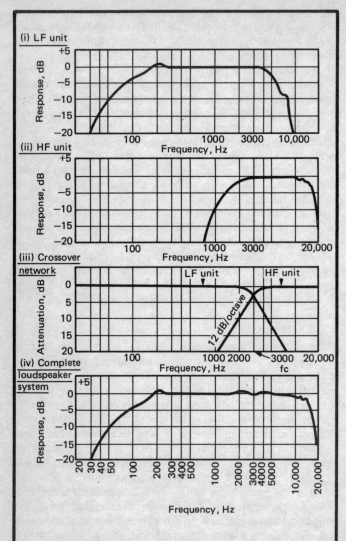

Fig. 5.15 HF and LF units combining
via crossover network

150

mouth distributing the sound into the open air. In effect the horn couples the diaphragm to a much larger volume of air than it could affect on its own. The net result is that horn systems have efficiencies as high as 50%. This is all summed up by the sketch in Fig.5.16(i).

A few simple equations can put us in touch with the design features. The shape of the horn may be conical, as with early megaphones, exponential or hyperbolic.(5/4.2.3.3) Exponential horns are probably in greatest use and because in electronics we already have experience with exponentials,(2/A4.2) the formula below are for this shape.

In common with many radiating devices, horns have a lower frequency of cut-off below which the transmission fails. Any horn is related to this frequency (f_{co}) by its *flare constant* which in a way shows how the horn spreads outwards, the constant is usually denoted by m . For a given cut-off frequency f_{co} :

$$m = \frac{4\pi f_{co}}{c} \tag{1}$$

(c is the velocity of sound waves in air) so having determined m , the horn equation follows:

$$A_x = A_t \cdot e^{mx} \tag{2}$$

where, as shown in the Figure, A_t is the cross-sectional area at the throat and A_x that at any distance x from the throat. e = 2.718. Equally, since $A = \pi d^2/4$ where d is the diameter,

$$d_x^2 = d_t^2 \cdot e^{mx} \tag{3}$$

A_t is known from the drive unit, A_x can therefore be calculated but we still have to determine the maximum value of x , that is the horn length, ℓ . It has been found by calculation and experiment that the mouth diameter should be at

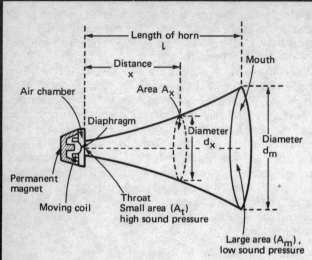

(i) Horn and driver

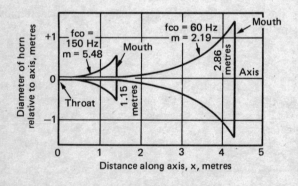
(ii) Exponential horns (calculated)

Fig. 5.16 Loudspeaker horns

least $\dfrac{\lambda}{2}$ at the cut-off frequency, f_{co} but note that there is little output at this frequency, the lowest *useful* frequency is some 25% higher. Then:

$$\text{mouth diameter, } d_m = \frac{\lambda}{2} = \frac{c}{2f_{co}}$$

and mouth area,

$$A_m = \pi \left(\frac{c}{4f_{co}} \right)^2 .$$

We do not get a very clear picture of what is happening by simply gazing at a string of formulae, let us develop these further to see what they mean — we cannot have too much practice in finding out for ourselves what formulae have to tell.

To see things on paper it is preferable to work in terms of diameters rather than areas, hence we have:

$$d_m = \frac{c}{2f_{co}}$$

from which, since

$$m = \frac{4\pi f_{co}}{c}, \quad m = \frac{2\pi}{d_m}$$

and since from equation (3)

$$e^{mx} = \frac{d_x^2}{d_t^2}$$

when $x = \ell$ then
$$e^{m\ell} = \frac{d_m^2}{d_t^2}$$

$$\therefore \; m \times \ell = \log_e\left(\frac{d_m}{d_t}\right)^2 = 2\log_e\left(\frac{d_m}{d_t}\right)$$

$$\therefore \; \ell = \frac{2\log_e\left(\dfrac{d_m}{d_t}\right)}{\dfrac{2\pi}{d_m}} = \frac{d_m\log_e\left(\dfrac{d_m}{d_t}\right)}{\pi}$$

and for, say, $f_{co} = 60$ Hz and a throat diameter of 2.5 cm:

$$d_m = \frac{344}{2 \times 60} = 2.87 \text{ metres}, \qquad m = \frac{2\pi}{d_m} = 2.19$$

and
$$\ell = \frac{2.87\log_e\dfrac{2.87}{0.025}}{\pi}$$

(a scientific calculator certainly makes life easier here), i.e.

$$\ell = 4.3 \text{ metres}.$$

A horn 4.3 metres long with a mouth diameter approaching 3 metres is very large and unwieldy (for $f_{co} = 30$ Hz the figures are 9.9 and 5.7 metres!), however the horn need not be straight, nor even circular, it can be *folded* as will be seen at the end of the next section.

The horn shape can be plotted from

$$d_x = \sqrt{d_t^2 \times e^{mx}}$$

and for example at 3 metres from the throat ($x = 3$), the diameter

$$d_x = \sqrt{(0.025)^2 \times e^{2.19 \times 3}} = 0.67 \text{ metres}$$

and the shape for this particular horn (i.e. throat diameter 2.5 cm, l.f. cut-off at 60 Hz) is shown in Fig.5.16(ii). It is repeated for a horn of the same throat diameter but $f_{co} = 150$ Hz so that we can see how raising f_c results in a smaller horn. The horn shapes we recognize in so many musical wind instruments. On Fig.5.16(ii) it is not possible to illustrate the horn for an h.f. unit with f_c say, 2000 Hz because it is too small for the scale, actually 3.4 cm long and of mouth diameter 8.6 cm (for the same throat diameter). The order of these figures indicates that horns used on h.f. units in cross-over systems are quite practical even in relatively small cabinets.

5.2.2.4 Enclosures

A complete loudspeaker comprises not only the conversion unit but also some kind of mounting such as a baffle board or the more aesthetically pleasing cabinet. These are acoustically necessary because the diaphragm creates sound waves at both its front and back and these waves are out of phase. We take first the example of a loudspeaker cone moving backwards thus creating a rarefaction in front of it and a compression behind. As shown in Fig.5.17(i), the higher pressure in the compression causes air to flow from the rear round the unit to "fill" the rarefaction at the front so tending to cancel out the radiation. When the cone moves forward the air flow is in the opposite direction. Because what we ultimately hear is the combination of two separate waves it is sometimes known as an *acoustic doublet*.

At low frequencies the cone movement is relatively slow, for example, at 50 Hz a movement over half a cycle takes 10 ms whereas the time at 10,000 Hz is a mere 0.05 ms. Cancellation of one wave by another takes time, hence because at low frequencies more time elapses while the wave is being set up, a greater cancellation is possible. On its own therefore a loudspeaker unit spoils its own low frequency performance. The wave emanating from the rear of the unit has therefore

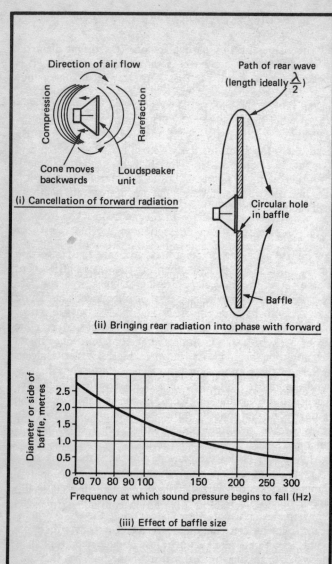

Direction of air flow

Compression

Rarefaction

Cone moves
backwards

Loudspeaker
unit

(i) Cancellation of forward radiation

Path of rear wave
(length ideally $\frac{\lambda}{2}$)

Circular hole
in baffle

Baffle

(ii) Bringing rear radiation into phase with forward

Diameter or side of
baffle, metres

Frequency at which sound pressure begins to fall (Hz)

(iii) Effect of baffle size

Fig. 5.17 Loudspeaker baffles

either to be brought into phase with that at the front or be removed altogether.

(i) The Flat Baffle

A *baffle*, while not truly an enclosure, is included here and described first because its method of bringing the rear and forward radiations more into phase is uncomplicated, it simply forces the rear wave to travel that much farther on its journey round to the front for its phase to have moved more into line. The baffle is a large plain board, most likely of wood or chipboard, with the loudspeaker unit mounted usually but not necessarily in the centre as shown in Fig.5.17(ii). Radiation from the rear of the cone now has to travel around the board so that, put simply, during the time that a compression from the rear reaches and mixes with the radiation at the front, the latter has changed from rarefaction to compression, hence there is no cancellation and in fact the two waves add. More technically, the rear wave must travel a distance of $\dfrac{\lambda}{2}$ to provide 180° phase change. Large baffles are therefore necessary to improve the low frequency performance, for example, at 40 Hz, $\dfrac{\lambda}{2}$ is over 4 metres so the baffle radius must be about 2 m or if the baffle is square, of about 4 m side. Of course there is nothing precise about this calculation for the region in which the two waves meet is ill-defined, nevertheless the principle holds good which is that by use of a large baffle the l.f. performance is greatly improved. A rough idea of the improvement due to baffle size can be gained from Fig.5.17(iii). Here we make the not unreasonable assumption that cut-off is beginning to make itself known when the sound pressure output from the loudspeaker has fallen from its "normal" level by 2 dB and is continuing to fall as frequency goes lower. The curve shown is a mixture of experiment and calculation but does at the deliberately chosen value of 2 dB exhibit a close relationship with one for a baffle diameter of $\frac{\lambda}{2}$. We must not be mislead by this graph into assuming that for a baffle of, say, 2.5 m diameter, frequencies below 70 Hz are not reproduced. 70 Hz is the frequency at which cut-off begins, not the frequency at which output is greatly reduced. Thus as frequency falls the loud-speaker output falls but not sharply, in fact at one octave lower (35 Hz) the sound pressure is only reduced by some 10 dB. This may not be as serious as it appears bearing in mind that bass controls on amplifiers (discussed in Sect. 6.1.2.1) can easily replace a loss of this magnitude.

(ii) The Folded Baffle
This is where a flat baffle begins to change into a cabinet for we simply imagine a square board with part folded over as shown in Fig. 5.18(i). It is then more popularly known as an open-back cabinet or open-box. The obvious objection to the flat baffle of sheer size is partly overcome but unfortunately a new problem arises, that of air column resonance (Sect. 3.2.1) for the box is in fact a tube closed at one end [Fig. 3.4(ii)]. Sounds near resonance create a booming reproduction because they are reproduced at higher level and since resonance occurs at $\frac{\lambda}{4}$, it is usually well within the lower frequency range. Television and some radio receivers provide examples of partly open-back cabinets, they need to be partly open for ventilation purposes.

(iii) The Closed Cabinet
This would seem to be the most logical approach to the problem, the removal of the rear wave altogether. It is easily carried out [Fig. 5.18(ii)] but other difficulties arise. With the cabinet closed, air disturbed by the rear of the cone is no longer free and so experiences pressure changes which cannot propagate. The enclosed air therefore adds to the stiffness of the cone which has the effect of raising the bass resonant frequency of the loudspeaker unit itself, a typical example might be from 40 to as high as 100 Hz. Clearly the smaller

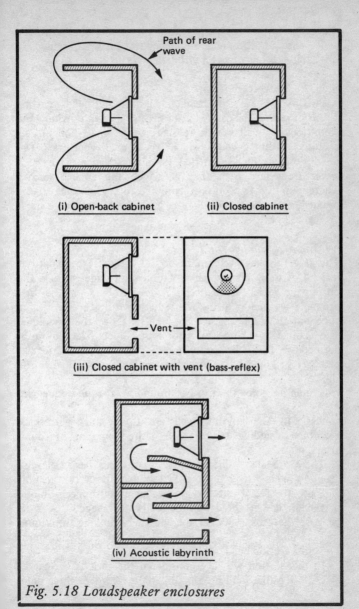

Fig. 5.18 Loudspeaker enclosures

the cabinet, the greater is the effect on the cone stiffness, hence raising the resonant frequency still further. Technically this cabinet is known as an *infinite baffle*.

(iv) The Bass-Reflex Cabinet

A shift back of resonant frequency is accomplished by cutting a *vent* or *port* in the cabinet as in Fig.5.18(iii) so providing the conditions for a Helmholtz resonator, that is, a volume of air enclosed within a vessel which has a small aperture. The dimensions are chosen so that the air resonance occurs at a lower frequency than that at which the cone resonates whereupon the two outputs combine to regain some of the original bass resonance ("reflex" indicates that the bass is "directed back").

(v) The Acoustic Labyrinth

This is a further development of the bass-reflex cabinet but instead of using air resonance within the cabinet to raise the output at lower frequencies, it reverts back to the first principle we met of using the radiation from the rear of the cone but shifted in phase. Fig.5.18(iv) shows the technique. The cabinet is divided into horizontal ducts (the labyrinth, defined in the dictionary as a "complicated, irregular passage) so that the total path length traversed before the rear radiation joins the front is around $\dfrac{\lambda}{2}$ for the frequencies concerned. The labyrinth walls are usually lined with thick absorbent wadding to create a desired amount of sound attenuation.

There are many different designs on this basis, the aim generally being to reduce the resonance peak of the unit and also to spread out the low frequency response. In one particular design the labyrinth walls are shaped so as to increase the cross-sectional area gradually running from loudspeaker unit to vent, the result being a low frequency horn. The cross-section need not be circular.

In this section we have considered the various enclosures as enclosing one loudspeaker unit only. Two or three units with cross-over networks may equally be built in, the number of different arrangements, all aiming at high-fidelity reproduction, is endless.

5.2.3 Electrostatic Loudspeakers

Although the electrostatic loudspeaker is much less frequently encountered than the moving coil, it is of interest because it is another example of a microphone principle in reverse, has high efficiency and is capable of a wide frequency response. In both the microphone and the loudspeaker the central unit has the form of a capacitor. As with the microphone we commence with the relationship $C = \dfrac{Q}{V}$ where C is the capacitance, Q the quantity of electricity held on the plates (the charge) and V is the voltage across the plates. Now a capacitor in its simplest form has two plates carrying opposite charges Q_1 and Q_2 and therefore there is a force of attraction between them. Coulomb (the French physicist) published the most fundamental law of electrostatics by stating that:

$$F = \frac{Q_1 \times Q_2}{4\pi \epsilon d^2} \, ,$$

ϵ is the permittivity for air[1/4.2.3] and d the distance between the plates. We can write

$$F \propto \frac{Q_1 \times Q_2}{d^2}$$

because these are the only factors concerning us here.

The electrostatic loudspeaker is simple in its basic conception, merely a rigid perforated metal plate as one electrode and very close to it a light metallic or metallized plastic diaphragm as illustrated in Fig. 5.19(i). The plates may be between half and one metre long. When charges appear on the plates due to an applied signal they are equal and of opposite polarity, the

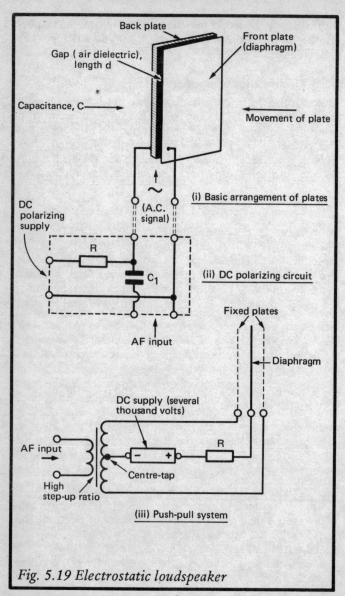

Fig. 5.19 Electrostatic loudspeaker

force F between the plates is therefore one of attraction and proportional to their magnitude ($F \propto Q_1 Q_2$). The force is also inversely proportional to the square of the gap length d, $\left(F \propto \dfrac{1}{d^2} \right)$, so for as high a force as possible the gap must be small. A gap of the order of 0.025 mm is usual and so that the diaphragm is free to move under the influence of the force F and yet not come into contact with the rigid backplate, it is held under tension. An advantage of this method of drive compared with that of the moving coil type is that the force acts evenly over the whole area of the diaphragm rather than being applied around the centre only.

If an a.c. signal is applied as shown in the Figure, whatever its polarity, the plates are oppositely charged and therefore attract, the diaphragm consequently moving inwards. The effect is that one cycle of input results in two pulses of rarefaction instead of one of rarefaction followed by one of compression, there is distortion with a doubling of frequency. If however a continuous d.c. potential is used to provide a steady charge on the plates and therefore a steady inward shift of the diaphragm, frequency doubling is prevented. This follows because on one half-cycle of the input waveform the steady charge is increased, pulling the diaphragm further inwards whereas on the other half-cycle it is decreased so allowing the diaphragm to move outwards. There still remains however, a problem of non-linearity. Assume the input signal is in such a direction as to pull the diaphragm inwards. This is mainly a linear move since $F \propto Q$. However, capacitance is thereby increased because the thickness of the dielectric is reduced so since $Q = CV$, Q is also increased, hence F increases so bringing in a second order effect, resulting in non-linearity. Prevention of this is by stopping Q from increasing owing to the capacitance change by use of a high-value resistor, R, in series with the d.c. polarizing supply. This maintains Q constant by severely curtailing electron flow. The circuit is shown in Fig. 5.19(ii) in which capacitor C_1 is used to block the d.c. from the a.f. input. The time constant $C_1 R$ must be long compared with the longest signal period (i.e. at the lowest frequency) otherwise Q will change appreciably during one cycle of the signal. Typically C_1 might have a value of 1500 pF for a unit with good l.f. response so R needs to be at least 10 MΩ for $C_1 R$ to be 15 ms, reasonably long compared with, say, the 10 ms period at 100 Hz.

Push-pull amplifiers[3/3.2.5] are noted for their ability to reduce spurious harmonics and push-pull working within an electrostatic loudspeaker is equally worthwhile. In this more advanced type the fixed plates are perforated or are of metal gauze while the conductive diaphragm is located between them. An arrangement with d.c. polarization is shown in Fig. 5.19(iii). With no input signal the two fixed plates are equally negative compared with the diaphragm, the forces on it are therefore equal but being in opposition, they balance out. Now when the transformer secondary adds an equal voltage in series with each plate but aiding for one and opposing for the other, the plate which becomes less negative reduces its attractive force on the diaphragm whereas the force for the more negative plate increases, hence push-pull action. This technique results in total harmonic distortion of very much below 1%.

Although an electrostatic loudspeaker is a masterpiece of precision engineering because of the extremely small gaps between the plates and high potentials involved, very successful models have been produced with frequency responses within ±5 dB from about 50 Hz up to some 20 kHz with in addition good response on transients because of the relatively low mass of the diaphragm.

Electrostatic units have a high impedance as might be expected, it is also almost purely capacitive. Special drive arrangements may consequently be needed in the power amplifier so that the unit is fed with a sufficiently high signal voltage. Fig. 5.19(iii) shows a step-up a.f. transformer as one solution.

5.3 EARPHONES

These are loudspeakers in miniature, they exist singly when, for example, used in telephones or hearing aids and in pairs coupled by a headband when they are then more popularly known as *headphones*. In telephone systems the single earphone of the *handset* is usually known as a *receiver*. It is not simply a case of making the item smaller however, some design features need modification, for a loudspeaker works into the open air whereas an earphone is coupled to the small volume of the ear via a peculiarly shaped and most unaccommodating entrance. Several differences become apparent: (i) the compliance of the air in the ear canal is equivalent to an added acoustic capacitance to the load, (ii) the seal between the earcap and the ear may be far from good, (iii) it has been established by experiment that we need a higher sound pressure of around 5 dB acting on our eardrums if the sound arises from an earphone compared with sound entering from the open air. All of this makes testing difficult. An *artificial ear* may be used which in one design consists of a cavity within a brass block specially shaped and slotted to simulate the acoustic properties of the ear canal. The receiver under test is coupled to this block and a tiny microphone at the remote end of the cavity represents the inner ear mechanism and measures the sound pressure. Crude perhaps, but it is all we have, nevertheless quite effective especially when the relationship between sound pressure as measured by the microphone and loudness as would be perceived via a human ear, is known.

Thus, although earphones operate on some of the basic principles already discussed in this chapter, we must appreciate that other design features arise. Nevertheless, because we now understand the principles involved there is no gain in examining earphone construction in detail, nor do we profit from a constructional drawing if one can easily be imagined from the sketches which already exist of the appropriate microphone or loudspeaker elements.

Generally, earphones work on one of four principles as follows.

5.3.1 Variable Reluctance

In the earlier form the magnetic flux flows through the diaphragm, this is generally superseded by units in which the flux flows through a low reluctance armature which is mechanically linked to the diaphragm situated immediately behind the ear-cap. The principles of variable reluctance units operating as microphones are contained in Sect.5.1.4.3 and drawings of microphone units which are equally applicable to earphones are Fig.5.9(i) and (ii). Speech currents in the windings vary the steady flux in the magnetic circuit which includes either the diaphragm (i) or armature (ii), these components therefore vibrate accordingly and in (ii) the armature movement is impressed on the diaphragm via the connecting rod.

Two variants of interest because of their use in millions in telephone systems are:

(i) the UK *rocking-armature* telephone receiver, the principle is as in Fig.5.9(ii) except that as shown in Fig.5.20(i), the armature rocks on a pivot on the central limb of the magnet system[5/1.7.3] and drives a light, non-magnetic alloy diaphragm. In the drawing the plan is added to show the two *torsion members*, metal strips which are twisted (i.e. they undergo torsion) when the armature moves, they therefore restore the armature to balance while no signal is present. It is another example of a push-pull system;

(ii) the USA *ring-armature* telephone receiver in which the driving unit is in circular form with the light alloy diaphragm fixed to a ring-shaped armature as shown in Fig.5.20(ii). In this design the diaphragm is driven at its periphery instead of at the centre so unwanted modes of vibration are less likely to arise.

In both cases response is within ±5 dB from about 200 to nearly 4000 Hz, a response good enough for telephony but in no way high fidelity.

5.3.2 Moving Coil

The principles are amply covered in Sect.5.2.2 on the moving-coil loudspeaker. The better quality units of this type may have a frequency response covering the full audio range say

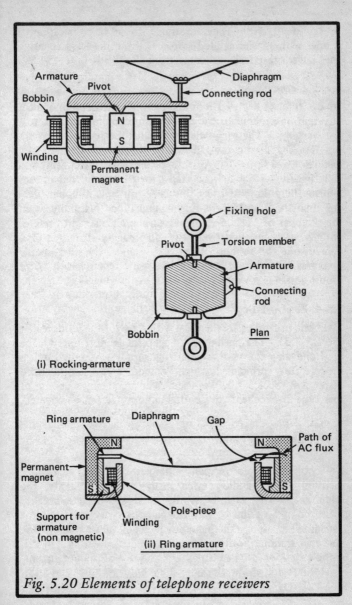

Fig. 5.20 Elements of telephone receivers

20 Hz to 20 kHz. They are usually found in pairs on head-phones with special padded earpieces. Even the less expensive ones have a response extending from about 30 Hz to 15 kHz.

5.3.3 Piezoelectric

We can refer to Sect.5.1.3 on the piezoelectric microphone for information on crystals and how pressure on one gives rise to a p.d. across it. The earphone principle is the reverse, that is, a p.d. applied across crystal faces causes it to distort, such distortion being used to drive a diaphragm. Fig.5.5(iii) and (iv) suggest typical drive systems for earphones and in fact loudspeakers do exist using this same principle although they are almost eclipsed by the moving coil type. Naturally when the sketches of Fig.5.5 represent earphones instead of micro-phones it is implicit that the diaphragms differ, in the microphone the function is to vibrate under the microscopic pressures of a sound wave, in the earphone a much more robust design is tolerable.

5.3.4 Electrostatic

The basic principles are identical with those of the loudspeaker (Sect.5.2.3). The push-pull arrangement is most likely to be used and an external d.c. polarizing supply is required unless the electret principle which embodies a self-polarized diaphragm (Sect.5.1.5.2) is used. The frequency response of this type is excellent, the entire audio range can be covered.

5.4 TELEPHONE SETS

No book on audio is really complete without at least a mention of an audio device many of us use daily, the telephone. This chapter is the most appropriate because from the audio point of view the part with which we are most concerned as users is the handset which contains a microphone and an earphone both of which are discussed earlier. In this section we look at the elements of the rest of the instrument.

Hybrid transformers and their relationship with the

Wheatstone Bridge and its balance conditions are probably already well known to the reader or may have been studied in Book 5 (Sections 4.1.1.1 and 4.1.1.2). Fig.5.21(i) shows the standard telephone instrument in block diagram form with its hybrid transformer linking the handset to the telephone line when switch S is operated. Balancing of the line to the set should be as good as possible because a poor balance allows too much power from the microphone to reach the earphone (via s to r), this is called *sidetone*. When sidetone is high (low attenuation s to r because of poor balance Z_ℓ to Z_b), ambient noise picked up by the microphone mixes with the incoming signal as shown and degrades the receiving signal/noise ratio. Also, when talking we hear ourselves via the same path and when high sidetone makes this abnormally loud we tend to lower our voices, this could be equated to installing a lower efficiency microphone, hardly good telephone practice.

The hybrid transformer has a second function that of coupling the microphone and earphone to the line, not only for impedance matching but also so that the desired balance of power transfer is obtained, for example, it may be helpful to raise the sending level to the line at the expense of receiving loudness or vice versa. Also surprisingly enough for instruments which contain no amplification, telephones can sometimes be too loud on receiving especially when the line attenuation to the distant end is low, for example, on a connexion between two users on short lines to the same exchange. Automatic sensitivity reduction is therefore frequently built into the transmission circuit.

In operation the springs S close when the handset is lifted and indicate to the local exchange that a connexion is required. Details of the connexion then follow from the dial or push buttons, the connexion is accordingly set up and switched through. Subsequently restoration of the springs S indicates to the exchange that the call is completed whereupon the exchange clears the whole connexion down.

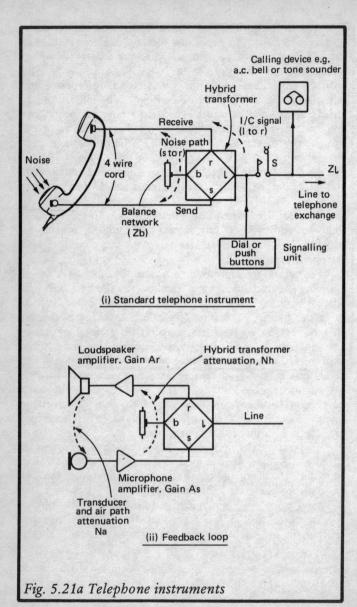

Fig. 5.21a Telephone instruments

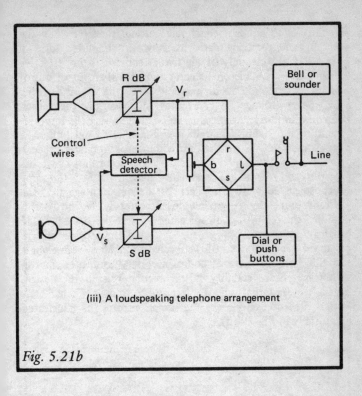

(iii) A loudspeaking telephone arrangement

Fig. 5.21b

5.4.1 Loudspeaking Telephones

We might start by looking upon a *loudspeaking* or *hands-free* telephone as in Fig.5.21(i) but dispensing with the handset and using instead a microphone at a short distance from the user and a small loudspeaker in place of the earphone. We must also add amplifiers because the speech sound pressure at the microphone is less and that from the loudspeaker greater because of the increased distance both transducers are now from the user. Immediately we run into the difficulty already met with hearing aids (Fig.2.4), that of acoustic instability for when in Fig.5.21(ii) $A_s + A_r$ exceeds $N_a + N_h$, unless a "press-to-talk" switch is used, the system is unstable and "howling" or whistling results. One way of increasing stability is to connect equal value attenuators[(5/3.2.3)] in the sending and receiving paths as shown in simplified form in Fig.5.21(iii), one only of the two attenuators being effective at any time. A speech detector switches them in or out of circuit as required. As an example, when the user is listening the speech voltage V_r exceeds V_s causing the speech detector to switch attenuator R out but S in. The improvement in the stability margin over Fig.5.21(ii) must then be equal to S dB. When the user commences to talk the speech detector finds that V_s exceeds V_r and accordingly switches S out and R in. The stability margin remains the same because S dB = R dB. There is the disadvantage that the distant user may be unable to interrupt while the user is holding the circuit but this is catered for if the voice is raised slightly so that V_r exceeds V_s and the attenuators are changed back in favour of the receive path. Many refinements to this basic idea are of course necessary to ensure smooth working. Since stability is the main hurdle in the design, again we find that smooth transducer frequency characteristics are necessary for the same reasons as illustrated pictorially in Fig.2.4(ii).

CHAPTER 6. AMPLIFIERS

Hardly ever in any audio project do we escape the need for amplification, loosely defined as creating a bigger signal from a small one. We of course know that there is more to amplification than that for a good amplifier must not only produce a bigger signal but this ought in all respects to resemble the input signal except for amplitude. There are voltage amplifiers and power amplifiers handling both small signals and large, there are also ones with different amplifications at different frequencies and there are integrated circuit amplifiers. In this Chapter we look at some of the more important basic design features first and then at the various types of amplifiers built around them.

6.1 DESIGN FEATURES

Most readers will be conversant with the basic working of a transistor amplifier[(3/3.2)] and also that

(i) *non-linearity distortion* arises whenever an output/input characteristic is not straight and that harmonic distortion[(2/1.4.3)] is a part of this;

(ii) *amplitude/frequency distortion* arises whenever components are encountered with characteristics which change with frequency;

(iii) *noise* is inherent in an amplifier as it is in all resistance, if followed by high amplification a poor signal/noise ratio results.

Generally therefore special arrangements are required to reduce these undesirable influences and usually the aim is for a flat frequency response. However, sometimes there is a deliberate move away from the flat response as in the bass and treble tone controls and for the special requirements of recording systems. To get to know these better we also look at the elementary design features of frequency dependent networks.

173

6.1.1 Negative Feedback

Of benefit to any amplifier is negative feedback. Most readers will have some acquaintance with the principles but a little revision never comes amiss especially in view of the growing use of operational amplifiers which rely on it.

Negative feedback (n.f.b.) occurs when a small part of the output is fed back into the input to go round again as occurs in the simple arrangement in Fig.6.1(i). The voltage gain of the amplifier without feedback is represented by A, β is the fraction of the output fed back. If the voltage at the amplifier input is v_1 then the output voltage $v_2 = Av_1$ and fed back is βv_2, therefore

$$v_s + \beta v_2 = v_1$$

where v_s is the signal input voltage and $v_2 = A(v_s + \beta v_2)$ from which

$$v_2 = \frac{Av_s}{1 - \beta A}$$

and the *overall* gain with feedback

$$A' = \frac{v_2}{v_s} = \frac{A}{1 - \beta A}$$

which is the fundamental expression for n.f.b. and it is on the assumption that v_s is increased when feedback is effective to regain the original output voltage v_2. Alternatively the amplifier gain may be increased as compensation, for since

$$A' = \frac{A}{1 - \beta A} \, ,$$

then change in gain equals

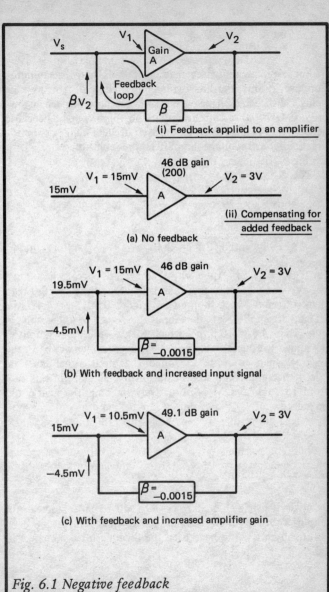

Fig. 6.1 Negative feedback

175

$$20 \log \frac{A'}{A} = 20 \log \frac{1}{1 - \beta A}.$$

Some practical figures may make this more meaningful. Suppose an amplifier has a *voltage* gain of 46 dB [here we are being a little slack in our use of decibel notation because we are not taking the amplifier input and output impedances into account — we can excuse ourselves because this is common practice where voltage amplifiers are concerned[5/2.1.1.1]], see Fig.6.1(ii).

Then

$$46 \,(\text{dB}) = 20 \log \frac{v_2}{v_1} \quad \therefore \quad 2.3 = \log \frac{v_2}{v_1}$$

$$\therefore \frac{v_2}{v_1} = \text{antilog } 2.3 = 200.$$

(a) If $v_s = 15$ mV, then $v_1 = 15$ mV and $v_2 = 15 \times 200$ mV = 3 V. Next suppose feedback is added with $\beta = -0.0015$, then voltage fed back to input = 3 V x −0.0015 = −4.5 mV which, if applied at the input in series would result in $v_1 = 15 - 4.5 = 10.5$ mV which with a 46 dB gain amplifier no longer produces 3 V output. Thus as shown at (b) in the Figure, when feedback is added the input signal must be raised to compensate for the fall in overall gain, in this case from 15 to 19.5 mV. Alternatively the amplifier gain could be raised. If v_s remains at 15 mV and v_2 at 3 V, $v_1 = 10.5$ mV and

$$\frac{v_2}{v_1} = \frac{3 \text{ V}}{10.5 \text{ mV}} = 285.7$$

or in decibels, 20 log 285.7 = 49.1 dB , that is, increase in amplifier gain required = 3.1 dB, as shown at (c).

Equally we could have used the formula for the change in gain,

$$20 \log \frac{1}{1 - \beta A} = 20 \log \frac{1}{1 - (0.0015 \times 200)}$$

$$= 20 \log 1.4286 = 3.1 \text{ dB}.$$

There may be a little uncertainty about the use of a minus sign with β. For *negative* feedback the signal returned to the input path must *oppose* the input signal and therefore usually has 180° difference (or near this value). The phase reversal required can be either in the amplifier itself in which the output signal is 180° out of phase with the input and we would write $-A$ in Fig.6.1(i) or alternatively within the feedback path in which again input and output are 180° out of phase and then β would be negative. Certainly we cannot have $-A$ and $-\beta$ together for this is positive feedback which results in instability. In practice when both A and β are complex the question of operation with no liability to instability is not easily answered and we ourselves would certainly be in danger of getting bogged down in a sea of complex algebra.(2/1.3.4) In the 1920's several people began to analyse stability in feedback systems, among these the one we should especially remember is Harry Nyquist, an American scientist who gave us the Nyquist Stability Theorem. We will not go further into this but the name of Nyquist often crops up in the study of electronics so it is useful to know with which branch he is associated.

It is worth remembering also that when feedback is applied at an angle other than 180°, A' increases, the full formula being

$$A' = \frac{A}{\sqrt{1 + |\beta A|^2 - 2\beta A \cos \phi}}$$

not a formula to go into deeply unless we owe ourselves a grudge, however, it is quoted because it tells us some important facts. Remember $|\beta A|$ is the *modulus* of the product of the two complex quantities(2/1.3.4) β and A. ϕ is the total phase shift of the amplifier and feedback circuit at the frequency under consideration. It is evident that at $\phi = 180°$, $\cos \phi = -1$ and A' is minimum, in fact for the example above where $\beta A = 0.3$, $A' = 153.8$ the same figure as would be obtained by calculation from $A' = A/(1 - \beta A)$. What is of most interest especially in view of the very high values of A obtained in operational amplifiers is the result when βA is considerably greater than 1. $| \beta A |^2$ becomes very high indeed compared with both the 1 and 2 $| \beta A | \cos \phi$ in the denominator of the expression for A' so this boils down to approximately $A/\sqrt{| \beta A |^2}$ or $1/\beta$, giving a gain with feedback which is independent of A (remember only when A is large) and substantially independent of ϕ so that stability is less of a problem.

Both distortion and noise which occur within an amplifier and appear at the output have with n.f.b. a fraction β applied to the input, these components are amplified and being in antiphase to the originals tend to cancel them. Again we realise that if ϕ is not 180°, cancellation is less, but now considering only the case of $\phi = 180°$, it can be shown that both distortion and noise are reduced by $1/(1 - \beta A)$(3/3.2.7.2) and if βA is very much greater than 1, by $1/\beta$. If v_d and v_n are the distortion and noise output voltages without feedback, then with feedback their output magnitude can be expressed as

$$\frac{v_d}{1 - \beta A} + \frac{v_n}{1 - \beta A} ,$$

again demonstrating that if βA is large, distortion and noise are much reduced by n.f.b.

It also follows that the frequency response of an amplifier improves with n.f.b. in that it is "flatter" and the bandwidth is increased as shown in Fig.6.2 for a hypothetical amplifier with and without feedback. The reasoning is simple and we need not revert to mathematics. If the gain with feedback at any frequency within the amplifier range is approximately equal to $1/\beta$ and β itself is made independent of frequency (e.g. by using a resistive network), then the gain is constant.

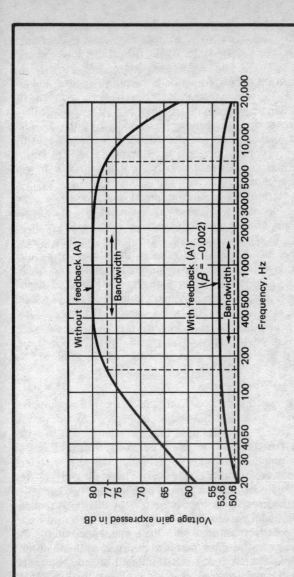

Fig. 6.2 Effect of negative feedback on gain and bandwidth

In Fig.6.2 this is demonstrated by the fact that with feedback the flat part of the characteristic has increased. However, when the amplifier gain falls at the low and high ends, βA falls and we can no longer rely on the $1/\beta$ approximation. Less feedback is therefore applied and the reduction in gain is smaller. Nevertheless even in the example shown which in fact does not have a particularly high value of β , the bandwidth as judged by the generally accepted method (between the -3 dB points relative to the mid-frequency gain) increases from about 150–7000 Hz for the amplifier without feedback to about 20–20,000 Hz with feedback as shown.

There is also an important practical point which arises from the fact that the gain reduction at the two ends in the original amplifier is due to reactive effects, especially those of shunt and series capacitance. With feedback, as frequency approaches the bottom or top ends, the phase shifts which arise within the amplifier bring instability nearer and in fact reduce the gain less than that shown in the Figure. This is a reminder that the simplified approach does not always tell the full story and at times the more complex formula quoted above for the gain with feedback must be used.

The Figure also demonstrates how through n.f.b. bandwidth can be traded for gain according to the value of β .

6.1.1.1 Feedback Modes

Feedback may be derived from the output of an amplifier proportional to either the output voltage or to the output current. It can be introduced into the input circuit in series or shunt (parallel). This gives rise to four main classes of feedback amplifier as shown in Fig.6.3. Several different descriptions are used for these, usually by two words relating to input and output and separated by a hyphen, two examples are "series-current", "voltage-current". A difficulty remains in recalling which word refers to input and which to output, thus we ourselves continue with the method adopted for the earlier books in the series, that is to mention input and output in all titles. Output is stated before input because this is the direction in which feedback goes.

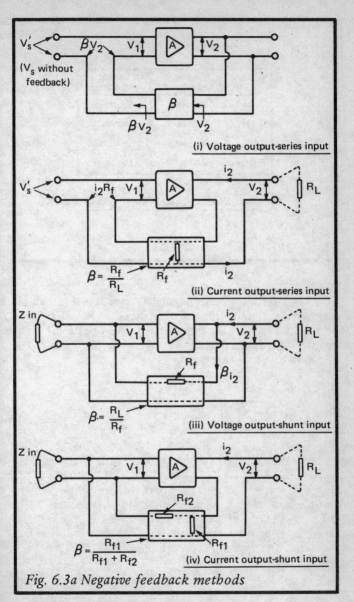

Fig. 6.3a Negative feedback methods

181

Method	Input impedance	Output impedance
(i)	Increases	Decreases
(ii)	Increases	Increases
(iii)	Decreases	Decreases
(iv)	Decreases	Increases

(v) Amplifier impedance changes with negative feedback

Fig. 6.3b

(i) *voltage output – series input* – the voltage is picked off the amplifier output [Fig.6.3(i)], reduced by the β network and fed back in series with the input to oppose the input signal. This is the system we have been discussing and it results in the general expression

$$A' = \frac{A}{1 - \beta A}$$

The current gain is unchanged for assuming that the input signal is raised to compensate for the addition of feedback, v_1 is unchanged, similarly with the current flowing into the amplifier and it must therefore have the same value throughout the input circuit. If the input current is unchanged when n.f.b. is applied yet a higher signal voltage is required, the inference is that the input impedance has increased. Let this be Z'_{in}, the raised dashes indicating that n.f.b. has been added. Then

$$Z'_{in} = \frac{v'_s}{i'_1} = \frac{v'_s}{i_1}$$

because $i'_1 = i_1$ as above. Note also that $v_s = v_1$.

$$\therefore \quad Z'_{in} = \frac{v'_s}{i_1} \times \frac{v_1}{v_1}$$

or by rearranging,

$$\frac{v_1}{i_1} \times \frac{v'_s}{v_1} \qquad \text{i.e. } Z_{in} \times \frac{v'_s}{v_1}$$

From Fig.6.3(i) $v_1 = v'_s + \beta v_2$ and since $v_2 = Av_1$,

$$v_1 = v'_s + \beta A v_1 \text{ from which}$$

$$\frac{v'_s}{v_1} = 1 - \beta A \quad \therefore \; Z'_{in} = Z_{in}(1 - \beta A).$$

Thus with either β or A negative, $Z'_{in} > Z_{in}$, that is, the input impedance is increased with feedback.

The change in output impedance with feedback can be reasoned by considering a load to be connected to the output terminals. Current flows in the output circuit hence v_2 falls because of the voltage drop in the amplifier output impedance. Accordingly βv_2 falls giving less feedback which tends to increase v_2. Thus the fall in v_2 is less than it would be without n.f.b. indicating that the output impedance of the amplifier must have decreased. By calculating the output impedance without and with n.f.b. it can be shown that

$$Z'_{out} = \frac{Z_{out}}{1 - \beta A}$$

(A being the amplifier gain with no load).

We next examine the other three feedback methods before looking at some practical circuits. Analysis of each takes a similar approach to that for (i) above, so we look at them more briefly.

Proper analysis of the input and output impedance changes when n.f.b. is added, although following normal Ohm's Law and amplifier principles, is time consuming. This is avoided therefore by simply saying that if the feedback circuit is connected in series then the impedance of that path is increased, if added in shunt the impedance is decreased. This makes some sense from the principle governing series and parallel impedances and is a good aid to the memory, never forgetting however that it is merely an artful expedient and is not the true explanation.

(ii) *current output − series input* − again a feedback voltage is injected in series with the input circuit but in this case the control is derived from the current in the output circuit. The Figure shows that the output current flows through a resistor R_f (or other network) producing a feedback voltage $i_2 R_f$. The feedback voltage is therefore proportional to the current in the load. Thus

$$\beta v_2 = i_2 R_f \quad \text{and} \quad v_2 = i_2 R_L$$

$$\therefore \frac{\beta v_2}{v_2} = \frac{i_2 R_f}{i_2 R_L} \quad \therefore \beta = \frac{R_f}{R_L} \, .$$

For the same reason as in (i) above the input impedance is increased, the output impedance can also be shown to be increased.

(iii) *voltage output − shunt input* − this method derives the feedback from the output voltage and applies it directly across the input circuit. Consider a single feedback resistor R_f and let the current flowing in $R_f = \beta i_2$ where i_2 is the amplifier output current. Then

184

$$\beta i_2 = \frac{v_2}{R_f + Z_{in}} \quad .$$

Assuming that Z_{in} is small compared with R_f and can therefore be neglected,

$$\beta i_2 = \frac{v_2}{R_f} \qquad \text{Also } i_2 = \frac{v_2}{R_L}$$

$$\therefore \frac{\beta v_2}{R_L} = \frac{v_2}{R_f} \qquad \therefore \beta = \frac{R_L}{R_f} \quad .$$

It is evident that shunt applied feedback does not change the voltage gain of the amplifier because neither the input nor output voltages are changed as happens with the series arrangements [Fig.6.3(i) and (ii)]. Because current is added to the input circuit from the feedback network however, more flows with feedback to obtain the same output, indicating that the amplifier input impedance has decreased. By reasoning not unlike that already used above it will be found that in this case the output impedance also decreases.

(iv) *current output − shunt input* − this feeds a fraction of the output current into the input circuit in opposition to the signal input current. The output current flows through R_{f_1} and $(R_{f_2} + Z_{in})$ in parallel or simply through R_{f_1} and R_{f_2} in parallel when $R_{f_2} \gg Z_{in}$, as can be arranged. The fraction β flowing through R_{f_2} and hence fed back into the input circuit is therefore

$$\frac{R_{f_1}}{R_{f_1} + R_{f_2}} \quad . \qquad (1/3.4.5)$$

As shown for (iii) above the input impedance decreases with feedback and we can judge from the output circuit that the output impedance increases. The range of impedance changes with feedback is tabulated in (v) of the Figure.

6.1.1.2 Feedback Circuits

To see how the feedback methods of Fig.6.3 are realized in practice, a few examples follow. On examining an amplifier circuit containing n.f.b. it is not always easy without some practice to determine which type is in use especially when, as we will see below, there may be a mixture. The following circuits give us this practice but even if we still find difficulty they at least illustrate how n.f.b. is so much a part of the modern amplifier.

(i) *voltage output – series input* – Fig.6.4(i) shows the circuit of a 2-stage transistor amplifier in which the feedback path is quite clear. The design avoids the use of a coupling capacitor between the two stages and also the need of one in the feedback path by suitable resistive arrangements, e.g. the d.c. potential on the collector of T_1 provides the correct bias on the base of T_2. Bias for T_1 is picked off the T_2 emitter resistor chain, R_5 and R_6. C_2 minimizes a.c. feedback. The effective feedback link from output to input is via R_1 and R_2 in series, the feedback is applied to the emitter of T_1 with

$$\beta = \frac{R_2}{R_1 + R_2}$$

That the feedback is negative follows from the fact that both T_1 and T_2 introduce 180° phase shifts, for example, for an n.p.n. transistor, should the base be driven negatively, the collector current decreases and the collector moves positively. From T_1 base to T_2 collector there is therefore 360° (or 0°) phase change. This in-phase voltage (as far as the base of T_1 is concerned) is fed back to the emitter thus opposing the base voltage for considering the T_1 emitter-base junction, shifting the emitter potential in one direction relative to the common line is equivalent to the same shift in potential on the

186

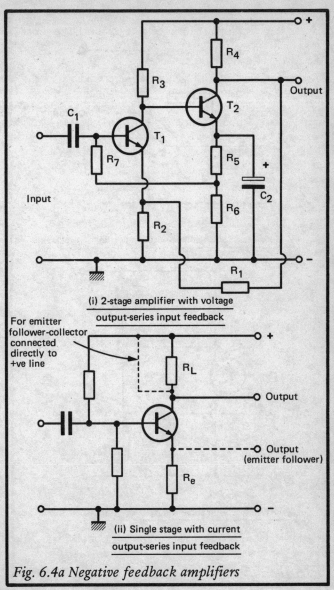

Input

(i) 2-stage amplifier with voltage output-series input feedback

For emitter follower-collector connected directly to +ve line

Output

Output (emitter follower)

(ii) Single stage with current output-series input feedback

Fig. 6.4a Negative feedback amplifiers

187

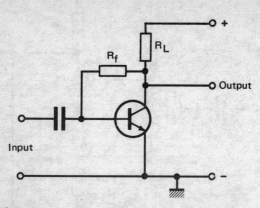

(iii) Single stage with voltage output-shunt input feedback

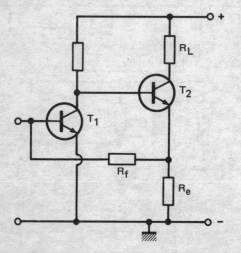

(iv) Two stage with current output-shunt input feedback

Fig. 6.4b

base but in the opposite direction. Hence there is 180° phase shift between emitter and base. Typically in such an amplifier, R_1 might have a value of about 15 kΩ and R_2, 1.5 kΩ, making

$$\beta = \frac{-1.5}{16.5} = -0.091 .$$

Approximate gain is therefore $1/\beta$ or about 21 dB.

In such an amplifier a high input impedance is often required especially when the amplifier serves a transducer which itself has a high impedance. Without feedback we might expect a gain of up to 100 per stage so if the input impedance is of the order of 1.5 kΩ without feedback then for the circuit shown

$$Z'_{in} = Z_{in}(1 - \beta A) = 1500[1 - (-0.091 \times 10^4)]$$

$$\approx 1366 \text{ k}\Omega$$

This is shunted by the resistors R_7 and R_6 in series. R_6 is negligible compared with R_7 which could be as high as 180 kΩ, thus the working input impedance is

$$\frac{1366 \times 180}{1366 + 180} \text{ k}\Omega \approx 160 \text{ k}\Omega ,$$

a usefully high value although hardly a match for an FET.(3/2.5)

(ii) *current output – series input* – the example is in Fig. 6.4(ii). It is very simple because by merely omitting the emitter by-pass capacitor of a normal transistor or FET stage, the emitter resistor R_e acquires a voltage across itself which is proportional to the emitter current and therefore approximately to the load current. This voltage is already in series in the emitter-base circuit and is in opposition to that on the base as we have seen above in the circuit in (i) which incidently also has some of this type of feedback built in because it has no emitter by-pass capacitor.

Reducing R_L to zero and taking the output from the top of R_e as shown dotted in the Figure, gives an interesting example of $\beta = 1$ because all the voltage developed by the load current is effective in providing feedback. The circuit is that of an *emitter follower*, so called because the output voltage "follows" the input voltage but with no amplifier gain, useless as an amplifier therefore but frequently used instead for impedance transformation from a very high to a very low value.

(iii) *voltage output – shunt input* – a circuit perhaps recognized more for d.c. stabilization is shown in Fig.6.4(iii). The output voltage is derived from the flow of the load current through R_L and this drives a current through R_f into the base circuit. Being opposite in phase it is subtracted from the base current. The circuit assumes that d.c. biassing is also correctly applied, if not then a capacitor is added in series with R_f to block d.c. and separate biassing arrangements made.

(iv) *current output – shunt input* – the emitter current of T_2 in Fig.6.4(iv) is proportional to the load current, it is therefore in order to pick off the voltage across R_e as a representation of it. This is fed via R_f across the emitter-base circuit of T_1. The current it produces is $180°$ out of phase with the signal base current, and is therefore subtracted from it. T_1 provides a phase reversal, but because the feed is from the emitter, T_2 does not.

6.1.1.3 Operational Amplifiers

So called because they were originally developed as analogue amplifiers for carrying out arithmetical *operations* in computers, this type in integrated circuit form is in such constant use that large scale production has reduced its price to a ridiculously low figure compared with similar circuits built from conventional components. For example commonly used operational amplifiers may contain some 20 transistors, half as many resistors with capacitors and/or diodes and all for less than the cost of a single small mica capacitor. We are considering the operational amplifier under the heading of negative feedback because in practically all cases it is employed as a feedback amplifier.

As an IC it is tiny, usually contained within a "flat pack" around 5 mm square and less than 2 mm deep. Its main characteristics as a general purpose amplifier are:

(i) high gain which can be reduced by negative feedback to a working value;

(ii) high input impedance so that little signal current flows to create a voltage drop in the source or low input impedance with certain types of feedback;

(iii) low output impedance so that the output voltage varies little with different loads;

(iv) large bandwidth, a desirable feature in most amplifiers.

Practical operational amplifiers can have gains without feedback up to 200,000, input impedances of the order of 1 MΩ upwards and output impedances less than 100 Ω. Bandwidths are made ample for audio working by using negative feedback as we saw in Sect.6.1.1.

The internal circuit diagram of an operational amplifier is quite complicated as one might expect from the number of components. Circuits are published by manufacturers but only the foolhardy will try to explain them completely because many integration techniques are employed which make transistor stages look just what they are not. There is little to gain from analysing the circuit and it is better to accept the IC as a single device. It differs from the conventional amplifier mainly in that there are two input terminals marked + and $-$. The first (+) indicates that a signal applied to that terminal appears amplified at the output with zero phase shift, the minus sign indicates that the output is 180° out of phase. The signs are *not* polarities but simply refer to output in phase (+) or opposite phase ($-$).

Biassing of the input stages is arranged externally to the IC. To fully understand the special requirements of biassing we must first consider the type of output circuit which is likely to be employed, the complementary symmetry Class B.

Other types may be used but the principles are the same. This circuit is looked at in more detail in Sect.6.3.2 on power amplifiers so we will confine our present investigation to the biassing problem only. Fig.6.5(i) shows the elements of such an output stage. T_1 and T_2 are a matched pair of transistors consisting of one npn and one pnp. By suitable choice of resistors $R_1 - R_5$ the base of T_1 is held +ve to its emitter while the base of T_2 is held equally −ve to its emitter so that the total resistances of $(T_1$ and $R_3)$ and $(T_2$ and $R_4)$ are equal and the output terminal therefore assumes a d.c. potential half that of the supply, i.e. V/2. When a signal arrives at the output stage from the driver transistor (not shown) T_1 conducts but T_2 is off with positive half-cycles and vice versa on the negative half-cycles in the normal Class B manner.(3/3.2.5.1) It is important that neither transistor swings to saturation otherwise the waveform peaks are clipped. Now if the d.c. biassing is not properly balanced or something causes it to shift it is evident that one transistor works with less than V/2 and is more likely to run into saturation while the other has a supply higher than V/2 and would be less likely to saturate. Conditions for distortion are therefore arising dependent on the degree of bias shift. Now in an operational amplifier which is d.c. coupled throughout, tiny d.c. unbalances which are inevitable may be followed by sufficient amplification that they result in considerable distortion and in the extreme the output terminal can be locked to V+ or 0V making one transistor of the pair ineffective with rectification rather than amplification the result.

Usually operational amplifiers work with a centre-tapped power supply (e.g. +15/0/−15 volts) in which case the output terminal as in Fig.6.5(i) should be at 0 V. From what we have discussed above and from the fact that the power supply voltages may not be exactly equal, the output terminal is unlikely to be precisely at this level. It is important that it should be therefore external arrangements are necessary for each operational amplifier to achieve the balanced condition. The technique is to adjust by changing the d.c. potentials at the input for there only a very small change is required. The

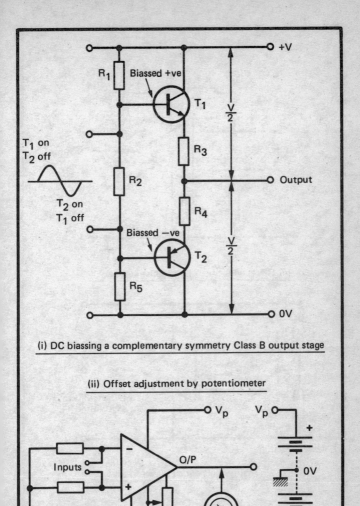

(i) DC biassing a complementary symmetry Class B output stage

(ii) Offset adjustment by potentiometer

Fig. 6.5a Biassing the operational amplifier

193

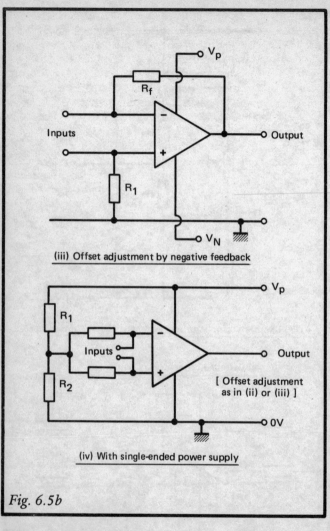

(iii) Offset adjustment by negative feedback

(iv) With single-ended power supply

[Offset adjustment
as in (ii) or (iii)]

Fig. 6.5b

difference of potential needed on the two input terminals to bring the output terminal to zero volts is known as the *input offset voltage* and the maximum value (a few mV) for any type of operational amplifier is quoted by the manufacturer. Fig.6.5(ii) shows perhaps the more obvious method of adjustment, offset correction pins are provided on the amplifier especially for this. The offset potentiometer (say about 10 kΩ) is simply adjusted for zero reading on the voltmeter, the latter is then removed but the potentiometer remains permanently wired into the circuit. A more automatic method which continually adjusts for changes in unbalances is by use of d.c. n.f.b. In this method as shown in Fig.6.5(iii), the + input is connected via a resistor R_1 to the 0 V (chassis) line and a feedback resistor R_f is connected between the output and − input terminals. Any departure from 0 V at the output is fed back via R_f to the − input which provides correction because potential changes on it in one direction result in magnified changes in the output in the opposite direction.

The centre-tapped power supply is not essential, a single-ended one can be employed as in Fig.6.5(iv). R_1 and R_2 are theoretically equal thus the input terminals are at $V_p/2$. Offset correction may be even more important because of a slight difference between the two resistors.

In normal use as inverting and non-inverting amplifiers, the basic circuits are shown in Fig.6.6.

(i) *non-inverting* − the input circuit is connected to the operational amplifier + input terminal and a.c. feedback of the voltage output − series input type is connected via R_f to the − input terminal. By analysis methods as used in Section 6.1.1.1 it can be shown that because the gain without feedback is very large, the circuit shown in the Figure has a gain

$$A' \approx \frac{R_f + R_1}{R_1}$$

and as would be expected from the type of feedback employed, with high input impedance (but shunted by R_2) and low output impedance.

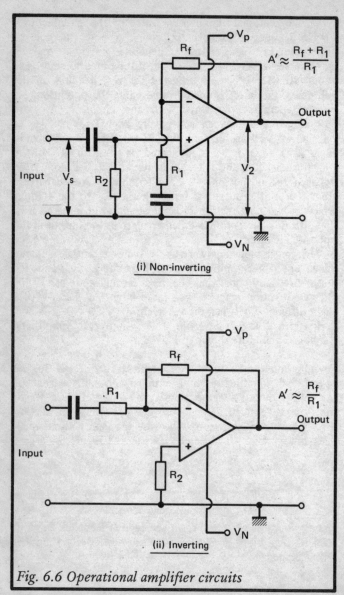

$$A' \approx \frac{R_f + R_1}{R_1}$$

(i) Non-inverting

$$A' \approx \frac{R_f}{R_1}$$

(ii) Inverting

Fig. 6.6 Operational amplifier circuits

196

(ii) *inverting* – both input and feedback are connected to the – input terminal, the + input being connected to the common line via R_2. Feedback is of the voltage output-shunt input type and considering the impedance of the source to be low then the gain can be shown as

$$A' \approx \frac{R_f}{R_1}.$$

Strictly this should be preceded by a minus sign because the gain is inverted (phase change of 180°). From Fig.6.3 we would expect the circuit to have low input impedance (but with R_1 in series) and low output impedance.

6.1.2 Frequency Dependent Networks

World-wide agreements on characteristics of recording and playback systems now tend to quote the requirements in terms of one or more *time constants*.[2/2] The understanding of this concept may at first seem a little elusive especially when we may have studied filters[5/3.3] for which time constants may not be mentioned. Simple filters however are based on capacitance and inductance together so involving resonance, on the other hand time constant techniques generally refer only to capacitance *or* inductance from which characteristics which change smoothly with frequency are derived. A little a.c. theory will put us into touch with how this happens and we will keep the arithmetic as simple as possible by choosing for our example a time constant, t of 504 μs.

Evidently in the circuit of Fig.6.7(i), as frequency rises the reactance of C falls ($X_C = 1/\omega C$ where $\omega = 2\pi f$) and therefore the output voltage V_R rises, pictorially shown by the phasor diagrams[2/1.3.4] in (ii) in which the phasor V is assumed constant. Thus the phasors V_R and V_C indicate the distribution of voltage in each case and also confirm that V_R increases and V_C falls as frequency rises. Our aim is to produce the attenuation/frequency characteristic of the network and fortunately the fact that V, V_R and V_C are not in phase is of no concern so we need not suffer the rigours of

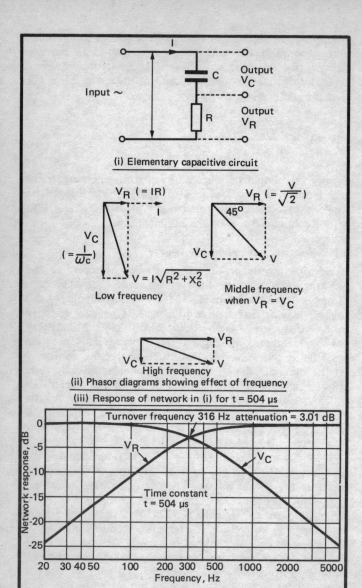

(i) Elementary capacitive circuit

(ii) Phasor diagrams showing effect of frequency

(iii) Response of network in (i) for t = 504 µs

Fig. 6.7 Frequency-dependent networks

198

operator j. In calculating the loss at any frequency we are comparing voltages so to avoid getting involved in negative dB values we can simply say that the number of decibels loss, $N = 20 \log(V/V_R)$.$^{(5/2.1.1)}$

Now

$$\frac{V}{V_R} = \frac{I\sqrt{R^2 + X_C^2}}{IR} = \frac{\sqrt{R^2 + \dfrac{1}{\omega^2 C^2}}}{R}$$

$$= \frac{\sqrt{\dfrac{\omega^2 C^2 R^2 + 1}{\omega^2 C^2}}}{R} = \sqrt{\frac{\omega^2 C^2 R^2 + 1}{\omega^2 C^2 R^2}}$$

$$= \sqrt{1 + \frac{1}{\omega^2 C^2 R^2}}$$

$$\therefore N = 20 \log \sqrt{1 + \frac{1}{\omega^2 C^2 R^2}} = 10 \log \left(1 + \frac{1}{\omega^2 C^2 R^2}\right)$$

[because $\log \sqrt{x} = \log x^{1/2} = \frac{1}{2} \log x$].

Now CR is the time constant t $\therefore C^2 R^2 = t^2$

$$\therefore \text{dB loss}, N = 10 \log \left(1 + \frac{1}{\omega^2 t^2}\right).$$

At the frequency at which $V_R = V_C$, $V = \sqrt{2} V_R$,

$$\therefore \frac{V}{V_R} = \sqrt{2} \quad \text{and} \quad N = 20 \log \sqrt{2} = 3.01 \text{ dB}.$$

Also
$$R = \frac{1}{\omega C} \quad \therefore \omega = \frac{1}{CR} = \frac{1}{t}.$$

This brings us to the *turnover frequency* f_t and since

$$\omega = 2\pi f, \quad f_t = \frac{1}{2\pi t}.$$

In the case we have chosen of $t = 504 \ \mu s$

$$f_t = \frac{10^6}{2\pi \times 504} = 316 \text{ Hz} .$$

A graph tells us so much more so we draw one for a range of frequencies above and below f_t, say 20–5000 Hz. Since $\omega^2 t^2 = 4\pi^2 f^2 t^2$, when $t = 504 \times 10^{-6}$, $\omega^2 t^2 = 4\pi^2 \times 504^2 \times 10^{-12} \times f^2 = f^2 \times 10^{-5}$ and

$$10 \log \left(1 + \frac{1}{\omega^2 C^2 R^2} \right) \quad \text{becomes} \quad 10 \log \left(1 + \frac{10^5}{f^2} \right)$$

more convenient to handle but only because of the deliberate choice of $t = 504 \ \mu s$. By substituting for f in this formula the graph of the network attenuation/frequency characteristic evolves, it is illustrated in Fig.6.7(iii) as network *response* which is simply the other way up and more illustrative of its effect. A second characteristic for V_C is included, calculated from the same principles, starting with

$$N = 20 \log \frac{V}{V_C} = 20 \log \frac{I \sqrt{R^2 + \dfrac{1}{\omega^2 C^2}}}{\dfrac{I}{\omega C}}$$

from which

$$N = 10 \log(1 + \omega^2 t^2) \quad \text{or} \quad 10 \log[1 + (f^2 \times 10^{-5})]$$

when $t = 504\ \mu s$.

The two curves in Fig.6.7(iii) show what can be done with the simplest of networks by suitable choice of turnover frequency. In this particular case the nearest preferred value components might be $C = 0.013\ \mu F$ (13 nF). $R = 39\ k\Omega$, producing a nominal time constant of 507 μs. Many other networks may be employed most with capacitance and resistance only but occasionally some of the more complex may include inductance. It is sufficient for us to have examined this simple series network only, design of parallel ones is not very different but mixtures of series and parallel with several different time constants need more advanced techniques. Nevertheless, hopefully this slight skirmish with the principles is sufficient for the subject to be no longer one of mystery.

6.1.2.1 Tone Control

We seem to spend much time talking about overall flat frequency responses and then deliberately introduce tone or bass and treble controls to do anything but keep the characteristic flat. From Chapter 4 however, rooms seem to have techniques of their own for modifying the output from a loudspeaker so some compensation may be necessary. Of greater consequence still is the fact that listeners have their own preferences as to what constitutes good tonal balance. In addition the loudspeaker itself may have a poor low frequency response so requiring *bass lift* or *boost*, while noise which is apparent more at high frequencies may be less objectionable with *treble cut*. Although these terms speak for themselves, they are shown graphically in Fig.6.8(i) in which the curves suggest typical maximum lifts and cuts.

The two factors which are most appropriate in defining a control characteristic are the turnover frequency [Fig. 6.7(iii)] and the slope, usually expressed as the number of decibels lift or cut per octave. There are two techniques of control:

(i) *Passive Tone Control* – bass and treble cut are obtained by simply attenuating the appropriate frequencies. In a passive

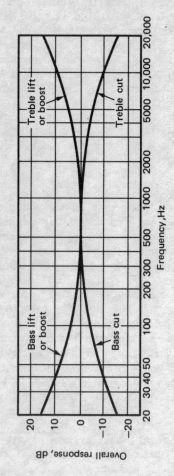

Frequency, Hz

Overall response, dB

Treble lift or boost

Treble cut

Bass lift or boost

Bass cut

(i) Typical tone control characteristics showing maximum lift and cut

Fig. 6.8 Passive tone control

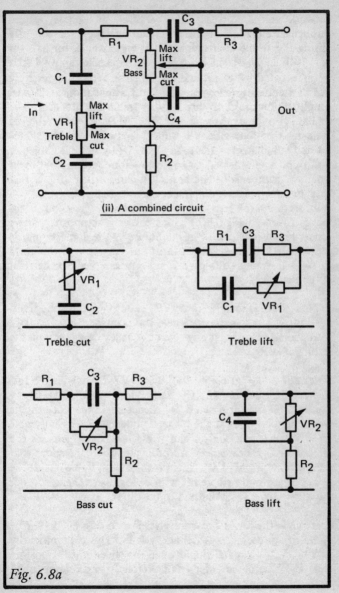

(ii) A combined circuit

Treble cut

Treble lift

Bass cut

Bass lift

Fig. 6.8a

203

network neither can actually be lifted so each is realized through attenuation of the other, amplification subsequently raising the whole characteristic. As an example, for bass lift of 15 dB at 20 Hz in Fig.6.8(i) all frequencies above 1 kHz are attenuated by 15 dB and those below 1 kHz progressively less as frequency falls, reaching 0 dB attenuation at 20 Hz. Amplification of 15 dB then restores treble and lifts bass. The similarity of curve *shape* in Fig.6.7(iii) to that required indicates that four series R and C networks could produce lift and cut characteristics as in Fig.6.8(i) by suitable choice of turnover frequencies. The control of four separate networks would be cumbersome and bearing in mind that if lift of either bass or treble is required, cut is not and vice versa, two controls only for bass and treble should be provided. This moves us on to Fig.6.8(ii) which shows a composite passive tone control network as might be used in a domestic music centre. It is no mean feat to sort out its action in this form so we will examine the parts separately as drawn to the right in the Figure.

treble cut arises mainly from VR_1 and C_2 across the trans-mission path, as frequency rises, the reactance of C_2 falls, with the attenuation at any given frequency dependent on the setting of VR_1 .

treble lift — C_1 in conjunction with VR_1 provides the lift. The falling reactance of C_1 as frequency rises progressively shunts the series arm $R_1 C_3 R_3$ hence the impedance of the whole network falls at the higher frequencies, so giving treble lift. VR_1 is the control and as its resistance increases the effect of C_1 decreases. Accordingly, when the slider is at the bottom end of VR_1 there is maximum treble cut, at the top end maximum treble lift with a progressive change over as the slider moves from one end to the other.

bass cut — the series capacitor gives rise to bass cut by virtue of its rising reactance as frequency falls. This is modified by part of VR_2 in parallel and when the slider is at the bottom end, C_3 is hardly affected and therefore has maximum effect, this therefore being the position for maximum cut.

bass lift — on the other hand bass is lifted when the shunt path of $C_4 R_2$ is effective by bypassing higher frequencies but not the lower. When the value of VR_2 which shunts C_4 is very high the capacitor has the greatest effect so creating maximum bass lift.

We have picked out the central features only for the sake of simplicity and ease of explanation. That each network shown does not function on its own and is also affected by the impedances of the input and output circuits is perhaps obvious. By suitable choice of component values a circuit as in Fig.6.8(ii) is capable of some 15 dB maximum lift or cut at the extremes of the audio range with none when the appropriate control (VR_1 or VR_2) is at the centre.

(ii) *Feedback Tone Control* — care has to be exercised in placing a passive tone control network within a system to avoid its loss driving the signal/noise ratio too low, as for example, by connecting a network to the input of a pre-amplifier. It is possible to use instead a network *within* the amplifier itself, in fact in its feedback path. This is equally effective and often more convenient especially with an operational amplifier which lends itself very well to this particular mode of use because of its need of an externally connected feedback network. If the feedback is made frequency sensitive in the opposite sense to that required of a passive network, the required degrees of tone control are produced. "Opposite sense" arises because feedback reduces gain so for treble cut, for example, the feedback network contains elements which *reduce* their attenuation at the higher frequencies to increase the feedback and reduce the amplifier gain. Such a network might be the one for treble *lift* in Fig.6.8(ii). With this in mind the complete network design follows the general principles as for the passive network and Fig.6.9 shows a basic IC arrangement. The same network might also be used connected to provide feedback from collector to base of a common emitter transistor stage.(3/2.2)

205

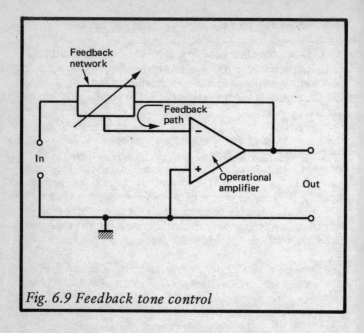

Fig. 6.9 Feedback tone control

6.1.2.2 Equalization

We have probably already met the *equalizer*[5/4.2.2.3] in earlier studies, a device which restores a frequency characteristic to its original shape. In Chapter 7 we shall find that in both disk and magnetic recording the amplitude/frequency characteristic is altered during the recording process either through deliberate manipulation or inherently. The characteristic has therefore to be restored within the playback chain. The equalizer which provides the correction usually takes one of two forms:

(i) a network having an attenuation/frequency characteristic complementary to the one needing correction as shown very simply in Fig.6.10(i) for a response which happens to rise with frequency. This is certainly not always the case but happens to be that way in some recording systems and the graph is not necessarily straight. (i) to (iii) in the Figure show equaliz-

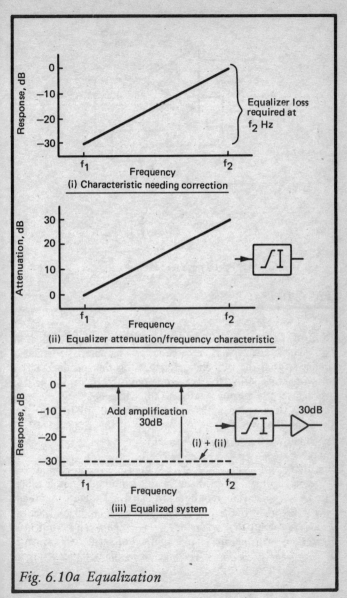

Fig. 6.10a Equalization

207

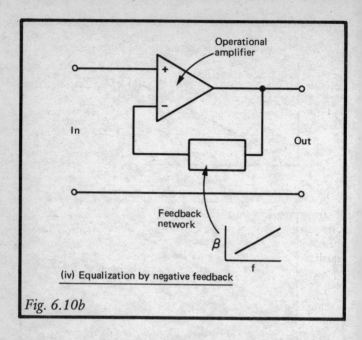

(iv) Equalization by negative feedback

Fig. 6.10b

ation of a *response* characteristic 30 dB lower at f_1 than at f_2 . The equalizer has an *attenuation* characteristic which is similar so that the two characteristics together produce a flat one, for considering the two ends only, at f_1 no equalizer loss leaves the response at -30 dB whereas at f_2 , 30 dB equalizer loss ·reduces the response from 0 dB to -30 dB. To bring the overall response back up to the original level of f_2 an amplifier gain of 30 dB is added.

(ii) instead of using a separate amplifier, the technique of frequency controlled feedback within an amplifier is feasible. For the equalization required in Fig.6.10(i) therefore a feed-back network is needed which increases the feedback fraction β at f_2 compared with that at f_1 . This can be put into figures by considering an amplifier of gain 10,000 with $\beta_1 = -0.002$ at f_1 . Then voltage gain with feedback at

$$f_1 = \frac{10,000}{1 - (-0.002 \times 10,000)} = \frac{10,000}{21} = 476 \ .$$

Now 30 dB represents a voltage ratio of antilog 30/20 = 31.62. ∴ voltage gain required at f_2 (30 dB less than at f_1) = 476/31.62 = 15. Then

$$15 = \frac{10,000}{(1 - \beta_2 \times 10,000)} \quad \text{from which } \beta = -0.067 \ ,$$

so in this particular case the feedback network must change β from 0.002 at f_1 to 0.067 at f_2. The shape of the β/frequency characteristic therefore follows that of the characteristic to be equalized. In this case we have not an equalizer as such but an amplifier with tailored frequency response instead. Fig.6.10(iv) shows the system using the high gain of an operational amplifier.

6.2 VOLTAGE AMPLIFIERS

In perhaps a rather negative way we might define a *voltage* amplifier as one which does not supply current as does a *power* amplifier, that is, it is driven by a voltage and our interest is only in the output voltage. Generally in audio work inputs are of the order of mV with outputs at a level of mV or a few volts.

So far we have experience of straightforward discrete component transistor amplifiers[3/3.2.4] and of operational amplifiers (from Sect.6.1.1.3). To advance a further step in understanding more specialized amplifiers and at the same time meet a type of circuit favoured in analogue IC's, especially operational amplifiers, we turn our attention to the *differential amplifier*, also known as a *balanced amplifier*, *emitter coupled pair* or *long-tailed pair*.

6.2.1 The Differential Amplifier

The essential features are illustrated by Fig.6.11 in which T_1 and T_2 are reasonably matched and $R_1 = R_2$. R_E is an emitter resistor common to both transistors. Suppose input 2 is held at the potential required to bias T_2 correctly and that a signal is applied to input 1. Positive swings of the signal cause the collector current of T_1 to rise (with npn transistors), hence its collector voltage V_1 falls. The rise in current in R_E swings T_2 emitter positive, so reducing both T_2 bias and its collector current and resulting in a rise in collector voltage V_2. V_1 and V_2 are therefore 180° out of phase. However, R_E is of fairly high value compared with R_1 and R_2 so in effect it tends to keep the current through it relatively constant. Thus with a rise in current in T_1 and

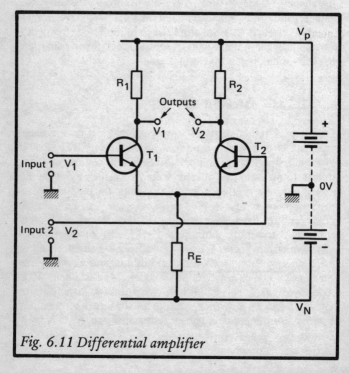

Fig. 6.11 Differential amplifier

fall in T_2 the voltage change across R_E is small so signals at the inputs do not change the standing emitter potentials significantly so minimizing any detrimental effect through the use of a common emitter resistor. Normally a signal is presented to both inputs at once [this is easiest to imagine as being from a centre-tapped transformer secondary winding as in Fig.6.12(iii)] with v_1 and v_2 in opposite phase (a *differential* input) then while the current in one transistor increases, that in the other decreases. Also while V_1 swings to one polarity, V_2 swings to the opposite and because the total output voltage is $V_1 - V_2$ and V_2 is negative relative to V_1, the net result is a voltage of $2V_1$ (or $2V_2$), assuming $V_1 = V_2$.

Common-mode or *longitudinal signals* (along, not across the wires) are those which affect both input terminals similarly. They are usually unwanted interference signals and are generated equally in both wires of a circuit pair (e.g. the leads from a microphone) by extraneous fields. Applied to the differential amplifier their input voltages are equal in phase, therefore the output voltages are also and the net output voltage is zero. Thus while handling a *transverse* signal (the signal voltage appears *across* the two input wires) properly, common-mode signals are rejected, a feature which makes the differential amplifier popular. However, so far we have assumed that T_1 has exactly the same characteristics as T_2 and that $R_1 = R_2$, this is perfection which we cannot have although we can approach it. We express this degree of perfection numerically through the *common-mode rejection ratio* (CMRR) which is the ratio between the differential voltage amplification (v_1 and v_2 in opposite phase) to the common-mode amplification (v_1 and v_2 in phase), usually in decibels.

Operational amplifiers usually employ the differential principle and for these the CMRR has a value of about 90—100 dB which is much better than would normally be obtained with discrete components. This superiority is due to the better matching of transistors within an IC in which T_1 and T_2 are likely to be adjacent. Also apart from the rejection of common-mode input signals, the differential amplifier has another worthwhile advantage. Temperature and power supply variations and *drift* (slow variations in d.c. conditions) all affect T_1 and T_2 equally hence cancel out at the output terminals.

6.3 POWER AMPLIFIERS

For the lower output powers, as with voltage amplifiers, the IC supersedes the discrete component circuit through its offer of much lower cost and size. For the larger output powers however the IC cannot yet compete. Power amplifiers have the problem of heat dissipation because not all the power generated goes to the load, that which does not becomes heat in the output transistors. With discrete components the heat is lost to the surrounding air by conduction from metal fins called *heat sinks*.[(3/2.3)] IC power output stages have more difficulty in conducting heat away but heat sinking is employed for the lower power outputs (say, up to some 20 watts or more) either through metal fins or by bolting the IC to a metal chassis. Thus for the larger output powers, output stages employing discrete transistors are more likely to be used. Evidently therefore the efficiency of an output stage is all important for the greater the efficiency in transferring power to the load, the less problem there is in getting rid of that remaining.

Efficiency can be expressed as $\dfrac{P_{ac}}{P_{dc}}$ where P_{ac} is the power into the load and P_{dc} the input power to the stage, i.e. supply voltage x total collector current. Also the collector dissipation $P_c = P_{dc} - P_{ac}$ so for, say, a 5 W output power from a 50% efficient output stage, 10 W are extracted from the d.c. power supply and 5 W dissipated in the transistor(s). This can give rise to a damaging amount of heat in the junction(s) if heat sinking is not employed. Because of the importance of efficiency therefore, we first look at the biassing modes available and the efficiencies they provide.

6.3.1 Modes of Operation

There are four main modes of operation called Class A, B, C and D. C is used in radio transmitters and D with switching circuits so do not concern us here. A and B are both used in amplifiers. In Class A a transistor is biassed approximately to the centre point of the straight portion of its transfer characteristic[3/3.2.4.2] as shown in Fig.6.12(i). A *quiescent* (motionless, from Latin for "quiet") collector current flows when there is no signal and the amplitude of the input signal is limited to avoid distortion through swinging into the non-linear parts of the characteristic. The fact that collector current flows throughout each cycle of the input waveform distinguishes the mode as Class A.

We can work out the efficiency of Class A as follows. Let the supply voltage $= V_{cc}$, then the maximum collector voltage swing (peak-to-peak) is from V_{cc} to 0, i.e. peak output voltage $= \dfrac{V_{cc}}{2}$. The r.m.s. output voltage is $\dfrac{1}{\sqrt{2}}$ times the peak value[2/1.3.3] and is therefore $\dfrac{V_{cc}}{2\sqrt{2}}$. Similarly the maximum r.m.s. output current is $\dfrac{I_{max}}{2\sqrt{2}}$ where I is the collector current. Therefore maximum output power is equal to

$$\frac{V_{cc}}{2\sqrt{2}} \times \frac{I_{max}}{2\sqrt{2}} = \frac{V_{cc} I_{max}}{8}.$$

Because Class A operates at the centre of the characteristic then

$$\text{quiescent current } I_q = \frac{I_{max}}{2}$$

hence maximum output power $= \dfrac{V_{cc} I_q}{4}$.

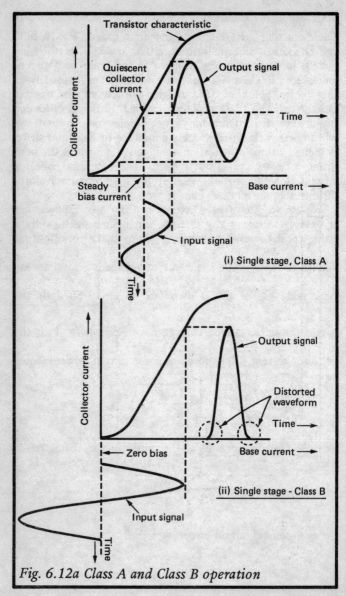

Fig. 6.12a Class A and Class B operation

214

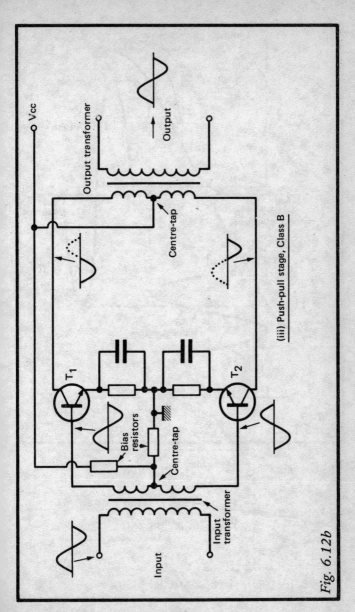

Vcc

Output transformer

Output

Centre-tap

(iii) Push-pull stage, Class B

T₁

Bias resistors

Centre-tap

T₂

Centre-tap

Input transformer

Input

Fig. 6.12b

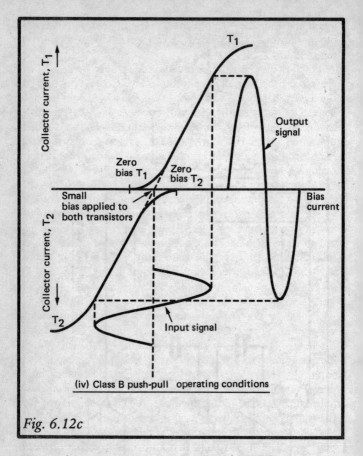

(iv) Class B push-pull operating conditions

Fig. 6.12c

The d.c. input power $= V_{cc} I_q$, therefore

$$\text{Efficiency} = \frac{P_{ac}}{P_{dc}} = \frac{\text{output power}}{\text{dc input power}}$$

$$= \frac{V_{cc} I_q}{4} \times \frac{1}{V_{cc} I_q} = 25\% \text{ ,}$$

216

less if the transistor is not driven to the full extent. Thus 75% of the power is dissipated in the transistor and must be removed.

Class B on the other hand is biassed near to cut-off and therefore has a low quiescent current. Fig.6.12(ii) shows how ridiculous it would be to use a single stage in Class B for instead of an amplifier it is of course a half-wave rectifier, the gross distortion of the waveform giving rise to many unwanted harmonics. Even the half-cycle which is amplified is distorted near the zero points (shown in the dotted circles) owing to the curvature of the characteristic at the bottom end. Apart from this, the fact that one half-cycle is reproduced reasonably faithfully leads to the *push-pull* amplifier where two transistors are used for the two half-cycles, each amplifying its own. A Class B push-pull circuit fast becoming out of date because of its need of transformers, but the one through which the principles are most easily revealed, is given in Fig.6.12(iii) with the operating curves in (iv).

The input transformer applies the signal at equal amplitudes but opposite phases to the matched transistors T_1 and T_2. As shown by the small sine waves in the diagram the collector voltages fall alternately, for T_1 on one half-cycle, for T_2 on the next. The collector currents combine in the centre-tapped output transformer primary to give a complete output waveform. In (iv) of the Figure the transistor characteristics are drawn compositely so that the effects of both half-cycles of input signal can be judged together. We can well imagine from (ii) that if zero bias were used on both transistors the distortion shown in the dotted circles would be present in the push-pull circuit. This is so and it is known as *crossover distortion* because it arises as the output wave crosses the axis. In Fig. 6.12(iv) however, a small bias is applied equally to T_1 and T_2 so that in fact there is a low quiescent current of value such that distortion is considerably less than with no bias. The Class B push-pull circuit therefore passes considerably less quiescent (collector) current compared with Class A and from this point of view alone, the circuit must be more efficient. Such a low quiescent current drain makes the system preferable when the d.c. power is derived from a primary cell with its relatively high cost. By ignoring the small quiescent current and considering a single transistor we can calculate the theoretical efficiency of Class B as follows:

For one transistor the collector current is simply a series of half sine waves, the r.m.s. value of the current being $\dfrac{I_{max}}{\sqrt{2}}$ and if the maximum collector voltage swing is considered to be between V_{cc} and zero then

$$\text{power into load} = \frac{I_{max}}{\sqrt{2}} \times \frac{V_{cc}}{\sqrt{2}}.$$

The average or mean value of the current for a sine wave is $\dfrac{2}{\pi} \times I_{max}$ (2/1.3.2).

$$\therefore \text{power taken from supply} = V_{cc} \times \frac{2}{\pi} I_{max}$$

$$\therefore \text{Efficiency} = \frac{\text{output power}}{\text{d.c. input power}}$$

$$= \frac{I_{max}}{\sqrt{2}} \times \frac{V_{cc}}{\sqrt{2}} \times \frac{\pi}{2V_{cc} I_{max}} = \frac{\pi}{4}$$

i.e. nearly 80%.

This is the absolute maximum, because a little quiescent current flows the practical efficiency is less but much superior to that in Class A. However, if efficiency is not a major consideration, the transistors in Fig.6.12(iii) can be biassed in Class A or somewhere between Class A and Class B, known appropriately as Class AB. Distortion is then lower than for Class B because curves in the transfer characteristics are avoided altogether.

6.3.2 The Complementary-Symmetry Class B Amplifier

As the capabilities of transistors developed, this type of circuit arose from the natural desire to rid power amplifiers of transformers which are bulky and expensive items compared with others almost disappearing from sight and of shrinking cost. The circuit is built around a matched pair of npn/pnp power transistors as labelled T_1 and T_2 in Fig. 6.13. These two transistors work in Class B with a small

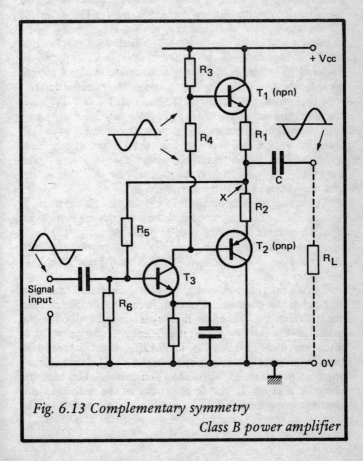

Fig. 6.13 Complementary symmetry

Class B power amplifier

quiescent current to minimize crossover distortion. Then assuming that the steady bias on each results in the same collector current, they have similar collector-emitter resistances and with the emitter resistors R_1 and R_2 equal, the junction between them (point X) is at half the supply voltage, i.e. $\dfrac{V_{cc}}{2}$. Across the d.c. supply there is a resistive chain comprising R_3 , R_4 , T_3 and its emitter resistor, moving up the chain the potentials are rising positively. The values of these components are such that a +ve bias is applied to T_1 , a negative bias to T_2 with respect to their emitters, the bias values being such as to maintain the quiescent current required. The bias is *forward* in both cases. A signal applied to the base of T_3 causes variations in its collector current, therefore voltage variations on the bases of T_1 and T_2. When the bases are taken positively, T_1 conducts and lowers its resistance, simultaneously T_2 is driven towards off so to high resistance. The net result is that the voltage at X rises towards $+V_{cc}$. Alternatively when the bases are taken negatively, T_1 goes to high resistance while T_2 falls to low and the voltage at X drops nearly to zero.

This swing in voltage at point X is applied via capacitor C to the load. Because complementary output transistors are used the circuit functions with in-phase signals to the bases and a phase-splitting driver stage(3/3.2.5.1) is not required. In the Figure are small sine waves to indicate the relative voltage swings.

T_3 is a driver transistor also providing the special facility of stabilizing the potential at point X by means of d.c. feedback. The d.c. potential at X is arranged through the potential divider R_5, R_6 to bias T_3 . If the potential at X rises more +ve), T_3 collector current increases so creating a greater voltage drop across R_3, R_4 thus shifting the base potentials of T_1 and T_2 negatively. This increases the resistance of T_1 but decreases that of T_2 with the result that the potential at X is reduced, so compensating for the original change. Equally should the potential at X tend to fall, the action of T_3 would be to raise it. Thus T_3

stabilizes the voltage at X at $V_{cc}/2$ volts.

In most cases the load R_L is one or more loudspeakers usually of 4 or 8 Ω impedance.

6.3.3 Integrated Circuit Power Amplifiers

IC amplifiers are unlikely to accomplish the whole job without a little outside help. Some components are too bulky to be accommodated, others must be outside for the purpose of adjustment. As a single example Fig.6.14 shows the components needed in addition to the IC for a typical small Class B amplifier of about 2 W output with 9 V supply and a total quiescent current of 4 mA. Such an amplifier would have an ample frequency response for the audio range, for example, within 3 dB from 20 Hz to 20 kHz. In Fig.6.14:

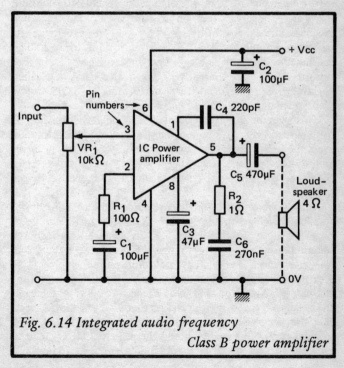

Fig. 6.14 Integrated audio frequency

Class B power amplifier

221

VR_1 is a gain control with R_1 and C_1 an additional arrangement setting the IC amplifier gain at a predetermined level

C_2 and C_3 are decoupling capacitors

C_4 is added externally to adjust the gain/frequency characteristic as required

C_5 blocks d.c. from the loudspeaker

R_2, C_6 may be required for stability reasons, e.g. to reduce the gain at frequencies well above audio.

It must be emphasized that every type of IC has its own individual requirements of external components. These are usually stated quite clearly by the manufacturer.

CHAPTER 7: AUDIO RECORDING

This Chapter concerns itself with audio systems in which the intention is to store not the fleeting sound wave itself but some equivalence of it. Two systems known to all are the gramophone disk or record in which sound wave pressures are captured in the walls of the record groove and the more recently developed magnetic tape in which a less tangible magnetic resemblance is used. Both systems are capable of very high quality reproduction. However, as digital techniques slowly creep in on analogue ones the future may have some changes in store, we therefore also look in on the possibilities.

7.1 DISK RECORDING AND REPLAY

Whether we talk about *gramophones* or *phonographs*, we mean the same, the written voice (from Greek). In this case sound waves are eventually cajoled into writing their message not on paper but in the groove of a disk, of a form known to all. With the passage of time the grooves have become smaller and turntable speeds slower even though stereo has also been accommodated, we will not therefore look back but consider the *microgroove* disk only with its two turntable speeds of 45 and 33⅓ revolutions per minute (r.p.m.) and stereo recording.

Amid discussion of extremely small groove and stylus dimensions we might find of interest an item surprisingly large, the actual length of groove on one side of a 30 cm (12 inch), 33⅓ r.p.m. disk. The simple calculation required shows this to be more than half a kilometre (about one-third of a mile) and this is the length of groove a playback stylus experiences each 'time one side of the disk is played, little wonder therefore that the hardness of a sapphire or diamond is needed.

7.1.1 The Disk

The modern microgroove is V-shaped and runs spirally, with

practically no exceptions, from outside to inside of the disk. Fig.7.1(i) shows a cross-section with approximate dimensions. The disk surface between grooves is known as the *land* and one unmodulated groove plus adjacent land is shown occupying a little less than one-tenth of one millimetre. We will see later that the width of the land can vary.

Stereo reproduction requires two separate channels to be recorded and this is accomplished by cutting a signal on each groove wall, the inner wall (nearer the disk centre) carries the signal for the L channel and the outer wall that for the R channel. This is known as a 45°/45° system because the cutting stylus moves at an angle of 45° to the disk surface causing variations in the groove walls as shown in Fig.7.1(ii). The arrangement ensures that if, for example, the R-channel carries a signal the cutting stylus varies the relative position of the appropriate groove wall because it exerts a force at right angles to it. This movement of the stylus has no effect however on the opposite wall except to vary its depth because the cutter face moves along it, not against it. The Figure shows a single sine wave on the R-channel and how the stylus changes the groove, shifting the R-groove wall into and out of the land but not so affecting the L wall. By projecting downwards in (ii) of the Figure an idea of how the groove would appear under a microscope is given, the R wall undulates in sine fashion whereas the L wall is undisturbed. Of course, the cutter stylus cannot move in two directions at once so when signals arrive on both channels, the actual stylus direction and degree of movement is the vector resultant(2/A4.1) of the two. This is quite in order because a vector can always be resolved into two components at right angles and this is what happens on playback.

The disk is formed from a thermoplastic material (softens on heating) usually of vinyl resin with a filler.

7.1.2 Recording

From Fig.7.1(ii) we see the shape of the tip of the cutting stylus, it is of sapphire, synthetic ruby or diamond for hardness. It is in fact a pointed chisel fixed in the cutter which as we have seen above, moves it to left or right according to the

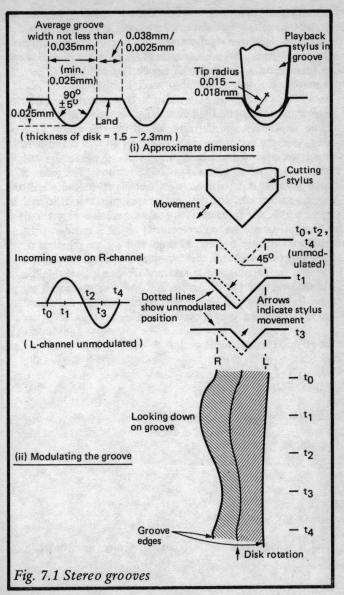

Average groove width not less than 0.035mm

0.038mm/0.0025mm

(min. 0.025mm)

90° ±5°

0.025mm

Land

(thickness of disk = 1.5 — 2.3mm)

Playback stylus in groove

Tip radius 0.015 — 0.018mm

(i) Approximate dimensions

Cutting stylus

Movement

t_0, t_2, t_4 (unmodulated)

Incoming wave on R-channel

45°

t_1

t_2 t_4

t_0 t_1 t_3

Dotted lines show unmodulated position

Arrows indicate stylus movement

(L-channel unmodulated)

t_3

R   L

— t_0

Looking down on groove

— t_1

(ii) Modulating the groove

— t_2

— t_3

— t_4

Groove edges

Disk rotation

Fig. 7.1 Stereo grooves

225

recording signals being fed in. The cutter is moved over the revolving disk so that the spiral track is formed and the rate at which it is moved determines the pitch between adjacent grooves. The stylus is heated, for example, by current through a winding on it, the heat aids the cutting process and ensures a smooth groove wall, thereby reducing noise.

From earlier studies we know that when a conductor moves in a magnetic field an e.m.f. is induced in it of magnitude depending on the *rate* of cutting the magnetic flux, hence on the *velocity* of the conductor.(1/5.3) If the conductor is coupled to a pickup stylus then the signal voltage output is proportional to the velocity of the stylus. The converse is also true, that is, for an electromagnetic cutting head the stylus velocity is proportional to the input voltage and if, irrespective of frequency, the input voltage is maintained constant, then the stylus velocity is also constant. Most cutter heads are of the electromagnetic type. Let us put a few practical figures to this. Assume a constant stylus velocity of 0.1 m/s (= 0.1 mm/ms) and frequencies of 100 and 10,000 Hz input at the same voltage. The maximum cut of the stylus into the land takes place between t_0 and t_1 of Fig.7.1(ii), i.e. over ¼ cycle, therefore

at 10,000 Hz time for ¼ cycle

$$= \frac{1}{4f} = 0.025 \text{ ms}$$

and recorded amplitude $= 0.0025$ mm;

at 100 Hz time for ¼ cycle

$$= \frac{1}{4f} = 2.5 \text{ ms}$$

and recorded amplitude $= 0.25$ mm,

and it is clear that stylus lateral movement is inversely proportional to frequency. Now at 100 Hz a lateral movement of 0.25 mm is unthinkable while as Fig.7.1(i) shows, the land, which is shared by two adjacent grooves has a width on average of about 0.04 mm. The problem is not solved by reducing the stylus velocity because this makes the high frequency recorded amplitude commensurate with disk imperfections and dust. Improvement is achieved instead by making the groove pitch and hence the land width variable. This is effected automatically in the cutting process, when low frequencies and/or high modulation levels are being recorded the pitch is increased (wider land) and vice versa, thus packing as many grooves as possible onto each disk. If the technique is exploited too much a groove will run into an adjacent one.

Further compensation is gained by attenuating low frequencies and accentuating the high ones during the recording process. The action on the high frequencies increases the lateral excursion and so reduces system noise. The process must be reversed on playback. Fortunately an internationally agreed standard recording characteristic exists, it is shown in Fig.7.2(i). Without doubt, the equation to the curve fills one with misgivings so we are not obliged to get involved except to remember the need for one and that through it approximately constant amplitude recording is achieved. However for readers with home computers or scientific calculators who wish to practise the art of getting the right answer sometimes, the complete formula as published is:

$$10 \log(1 + 4\pi^2 f^2 t_1^2) - 10 \log\left(1 + \frac{1}{4\pi^2 f^2 t_2^2}\right)$$

$$+ \ 10 \log\left(1 + \frac{1}{4\pi^2 f^2 t_3^2}\right) \quad \text{decibels,}$$

$t_1 = 75 \ \mu s$, $t_2 = 318 \ \mu s$, $t_3 = 3180 \ \mu s$. $t_1 - t_3$ are in fact time constants of frequency sensitive circuits (explained in detail in Sect.6.1.2) and information also published with the

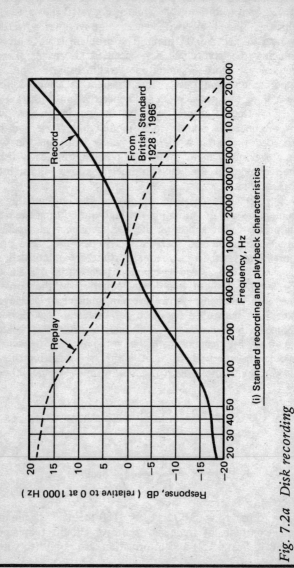

Fig. 7.2a Disk recording

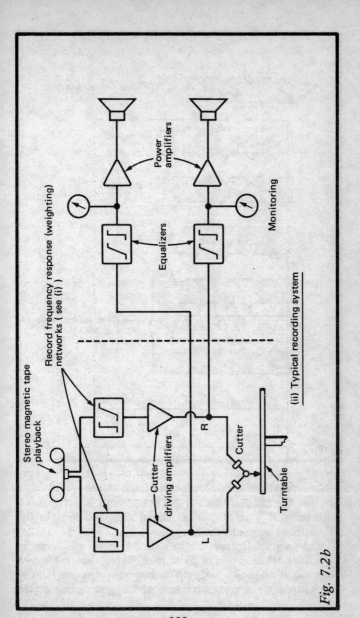

(ii) Typical recording system

Fig. 7.2b

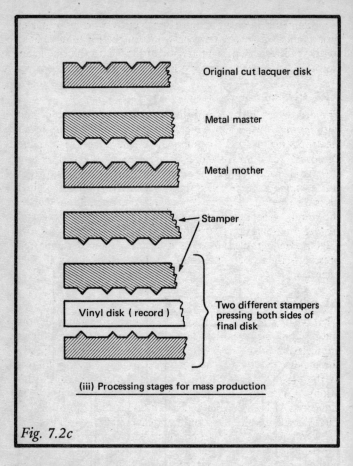

Original cut lacquer disk

Metal master

Metal mother

Stamper

Vinyl disk (record)

Two different stampers pressing both sides of final disk

(iii) Processing stages for mass production

Fig. 7.2c

recommendations assists designers of the circuits required to produce the full response curve. The formula is not quite so frightening if $4\pi^2 t^2$ is calculated first and held in the calculator or computer memory.

The essential features of the recording chain are shown in Fig.7.2(ii) which assumes as is most likely that the stereo recording is originally made up on magnetic tape (see later in this Chapter), the great advantage of doing so is that a tape recording can be changed until satisfactory whereas a master disk cannot. The *weighting* networks in the Figure are those to give the recording characteristic of Fig.7.2(i), the equalizers restore the frequency characteristic to normal for monitoring.

7.1.3 Disk Processing

The disk which is cut is most likely to be of aluminium coated with *lacquer*. When cut a *metal master* is then built up by an electroplating process so producing a "negative" in that the "grooves" stand up from the surface [see Fig.7.2(iii)]. From the metal master a *metal mother* is produced and again from this a *metal stamper*. The stamper is used to press the vinyl disks for playing. Some heat is used in this operation to soften the vinyl. The metal stamper wears during processing and further ones are pressed from the metal mother, not the original lacquer disk because it is too fragile for repeated pressings, the system therefore protects it by using it only once.

7.1.4 Reproduction

What we now understand about the recording process and how the resultant signal resides in the groove paves the way for an appreciation of the replay (or playback) requirements. Firstly the playback stylus plus its *cartridge*, usually together known as the *pickup*, has the job of sensing the groove width variations. This is most easily visualized if we consider an elementary electromagnetic system as sketched in Fig.7.3. Firstly let us recall how a magnet generates an e.m.f. in a coil of wire and for this we go right back to Faraday's early experiment.[1/5.3] In Fig.7.3(i) NS is a bar magnet which if pushed into or pulled out of the coil of wire generates an e.m.f. in it of magnitude depending on the velocity and strength of the magnet. The e.m.f. arises because the "lines of force" are cut by the coil turns and in this simplified theory it is assumed that the coil wires cut the flux at right angles. It follows therefore that if as in (ii), the magnet is moved at right angles to the original direction, no e.m.f. is generated because the coil wires move along the flux lines and do not cut them.

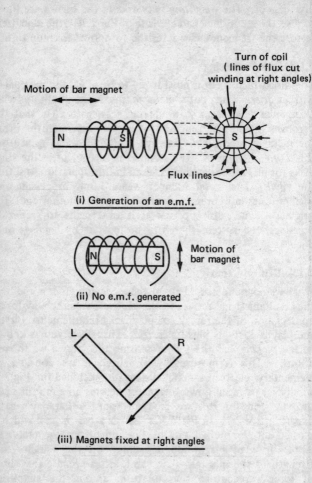

Motion of bar magnet

Turn of coil
(lines of flux cut
winding at right angles)

Flux lines

(i) Generation of an e.m.f.

Motion of
bar magnet

(ii) No e.m.f. generated

L   R

(iii) Magnets fixed at right angles

Fig. 7.3a Disc reproduction

232

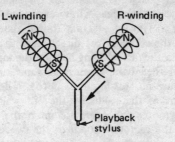

(iv) Generating L and R signals

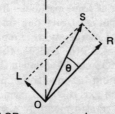

(v) OL and OR components give resultant motion OS of stylus

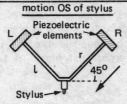

(vi) An arrangement using piezoelectric elements

Fig. 7.3b

Next suppose two bar magnets to be joined at right angles as in (iii). If they move in the direction of the arrow the left magnet moves "length on" whereas the right magnet moves "end on" and if this idea is now expanded to include coils and stylus as in (iv) we see that should the R wall of the groove move the stylus in the direction of the arrow an e.m.f. is generated in the R winding because the magnet movement is as in (i) of the Figure whereas no e.m.f. appears in the L winding because no flux is cut as in (ii). The principle holds for movement of the stylus by the L wall of the groove. When both grooves are operative upon the stylus it moves according to the vector resultant of the two motions but still generating the appropriate e.m.f.'s in the two coils. This is perhaps better explained pictorially. In (v) of the Figure are vectors[2/1.3.4] representing an amplitude (or velocity) movement of the stylus at some particular instant. There is a weak L-channel signal but a strong one on the R-channel. The resultant movement of the stylus is represented by the vector OS . This operating on the coils via the magnets has the large component OR (OS $\cos \theta$) acting in the direction of the axis of the R winding but the smaller component OL (S $\sin \theta$) in the direction of the axis of the L winding. In effect the two groove signals are transformed into a single movement of the stylus but because the two magnets are at right angles this single direction of movement is resolved back into the two original components. It is now evident that should both groove walls carry the same signal, equal L and R components result in a vertical movement of the stylus, a *mono* recording in fact.

7.1.4.1 *Pickups*

Clearly any mechanical to electrical transducer which is sensitive in one direction but not at 90° to this direction is suitable for a pick up element and this therefore includes piezoelectric types. Barium titanate ceramics are frequently employed, a typical arrangement is shown in Fig.7.3(vi). When the stylus moves, for example in the direction of the arrow, the R element is strained by the tension in its connecting arm (r) whereas the arm ℓ fixed to the L element mainly deflects. As is the case generally the voltage output of a piezoelectric element is higher than for an electromagnetic one, resulting in a much higher pickup impedance.

The mechanical design of pickups is a most complex process because two conflicting requirements have to be met, the stylus tip must remain in the groove yet its pressure on the disk must be low to avoid both disk and stylus wear. Advantage can be taken of resonance but there are also circumstances when it must be avoided. Thus the stylus mass, pick-up compliance and mechanical resistance must all be taken into account. These are quantities which are not easily measured but nowadays pickup systems have reached such a degree of mechanical sophistication that most of the difficulties have been overcome.

Magnetic cartridges exist in several forms, Fig.7.3(iv) shows the moving magnet principle and this is a type which is in use, others are variable reluctance and moving coil. Except for the latter their output is in the range from about 0.1 to 0.2 volts per m/s and if we assume an average stylus velocity to be around 0.05 m/s, the output voltage ranges from some 5–10 mV. Moving coil units have a much lower output, about 0.02 mV so because they have a low impedance need a step-up transformer to bring the voltage up to about 2 mV (100:1) and the impedance to around 50,000 Ω.

Piezo units have a much higher output voltage with a response of around 10 volts per m/s giving an average output of about 0.5 V. They require a high impedance load of 1–2 MΩ as we have already seen for microphones (Sect.5.1.3.1).

7.1.4.2 The Replay System

What is required is that the output from the reproducing loudspeaker should be a faithful copy of the original sound waves which were recorded. Many distortions creep in during the record/replay process and much effort is expended in high quality systems in reducing them, in addition Fig.7.2(i) shows how the recording characteristic is deliberately biassed against low frequencies and in favour of the high ones. If the frequency response of the system is otherwise flat then attenuation introduced into the record channel must be restored on replay, that is, in each channel an equalizer having a characteristic exactly complementary to the record characteristic [shown dotted on Fig.7.2(i)] is necessary for an uniform overall response. Such equalizers are shown in Fig.7.4 which illustrates the elements of a stereophonic reproducing system. Some notes on the techniques of equalizer design are given earlier in Sect.6.1.2.2. Note however that a piezo unit has an output which relates to amplitude of stylus movement rather than to its velocity, hence a different equalization is needed or special mechanical or electrical compensation may be included within the cartridge.

In the Figure L and R amplifiers raise the pickup voltage

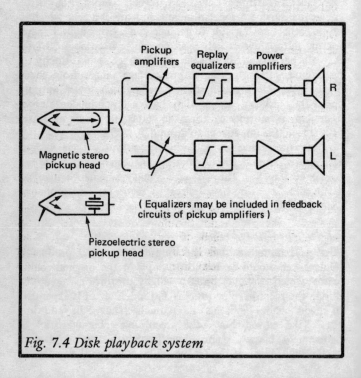

Fig. 7.4 Disk playback system

level and feed the equalizers via volume controls for adjusting the loudspeaker outputs. Further power amplification follows for driving the loudspeakers. Playback systems can have an overall frequency response varying only a few decibels from 20 to 20,000 Hz. This is checked by using a special disk on which is recorded at constant velocity a tone gliding over the audio frequency range.

7.1.4.3 Channel Separation

We probably know this title better as *crosstalk*, that is, the sound heard at the end of a channel resulting from interference from another channel. In stereo playback this simply refers to the degree to which L signals are heard on the R channel and vice versa. In the ideal system no crosstalk exists but on reconsidering Fig.7.3(iv) for example, we may begin to have doubts for it would be perfection indeed if no e.m.f. whatsoever is generated in a winding when a magnet moves across (not through) the coil. Thus the separation between two stereo channels is no more than some 20–35 dB, at certain frequencies even less, meaning that the crosstalk signal heard on a channel is 20–35 dB lower than the other channel signal. If a telephone listener heard crosstalk at 20 dB or even 35 dB he or she would complain bitterly for it would be audible but faint and certainly intelligible. However, stereo playback requirements are not so exacting, the two channels usually carry very similar signals and even if one were silent it only results in its loudspeaker reproducing faintly the signal on the other. But the crosstalk signal may be distorted so it must not be allowed to get out of hand, an average cartridge should therefore have at least 20 dB separation over as much of the audio range as possible. Generally stereo crosstalk is least over the middle frequencies (larger number of decibels), worsening at the lower and higher ends of the range and by as much as 10 dB at the extremes.

7.2 MAGNETIC RECORDING AND REPLAY

Although magnetic and disk recording systems have the same overall aim. they are anything but the same in the way they copy the input signal. Cutting a disk and rediscovering the information is a purely mechanical affair, on the other hand magnetic recording is electronic and nothing is changed permanently. This brings several operational advantages which will be appreciated as this section progresses. Firstly we look at the basic components of a magnetic system and then brush up on the theory of magnetism so that we may be well aware of what goes on in the tape itself.

Fig.7.5 shows the elements of a magnetic recording and replay system. The input signal is amplified as necessary and supplied together with an a.c. *bias* to the *record head* over which the *tape* moves at constant speed by the action of the tape driver roller or *capstan* from the supply to the take-up reel. The signal is recorded magnetically along the tape. The record head follows an *erase* head which demagnetises the tape so clearing it of anything previously recorded. For replay the tape is rewound onto the supply reel and again moved at the same speed but now with the *replay* head effective. The magnetization on the tape induces a replica of the original signal voltage into the replay head, this voltage is amplified, *equalized* and then appears at the output terminals.

7.2.1 The B-H Loop

Nowadays the theory is that spinning electrons are responsible for the magnetic properties of certain materials and just as an atom possesses a neutral charge when its protons and electrons balance,[1/1.1] so also does it exhibit no magnetism when the number of electrons with "clockwise" (+) spins balances the number with "anticlockwise" (−) spins. The atoms of some materials however, especially iron, nickel and cobalt, have an imbalance of + and − spins of the electrons in the 3rd shell[3/1.1.3] and consequently the material has magnetic properties. Note that as a conductor the *valency* electrons of iron are in the 4th shell. The fact that, for example, atoms of

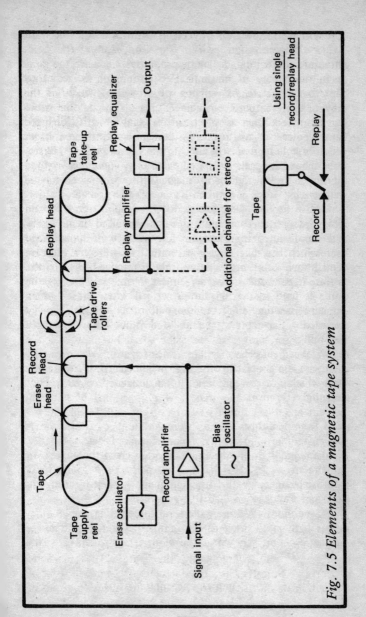

Fig. 7.5 Elements of a magnetic tape system

239

iron join with atoms of oxygen to form molecules of ferric oxide changes nothing within the iron atom, ferric oxide therefore also exhibits magnetic properties.

In the theory of magentization, groups of molecules are said to form *domains* (spheres of influence) which are our way of representing the tiniest volume of the material which can act as a magnet as we know one, that is, with North and South poles. To us therefore a domain is infinitesimal so we are only able to represent them crudely as in Fig.7.6(i) as tiny cube-shaped magnets polarized randomly, consequently overall the material exhibits no magnetism. Now when a weak magnetizing force is applied as in (ii) the domains tend to turn into the field direction as does a compass needle but the field lacks sufficient strength to fully align all of them. In (iii) a strong magnetizing force has fully aligned all domains, the material is then magnetically saturated. Most of the condition remains when the magnetizing force is removed, the material is then a *"permanent" magnet*, unless an opposing magnetizing force is used to destroy the effect. If when left alone the domains reassume their random pattern as in (i), it is not a permanent magnet and the material is useless for magnetic recording.

We recall that the magnetomotive force[1/5.2.1] (m.m.f.) is the magnetic equivalent of e.m.f. in the electrical circuit, it is that which creates a magnetic flux in a magnetic circuit. For a solenoid (air-core) the m.m.f. is proportional both to the current and the number of turns on the coil, i.e. m.m.f. $F = N \times I$ ampere-turns where N is the number of turns and I the current in amperes. For convenience, because F arises fundamentally from current, it is often given the dimension, ampere or A, i.e. $F = NI$ A . [If this unit is found to be confusing, one may feel more in touch with it by using At for "ampere-turns" instead.]

Magnetomotive force is the total force acting, the *intensity* or strength of the field is the m.m.f. per unit length of the magnetic circuit, known as the *magnetizing force*, H . Thus

$$H = \frac{F}{\ell} = \frac{NI}{\ell} \text{ amperes per metre (A/m)}$$

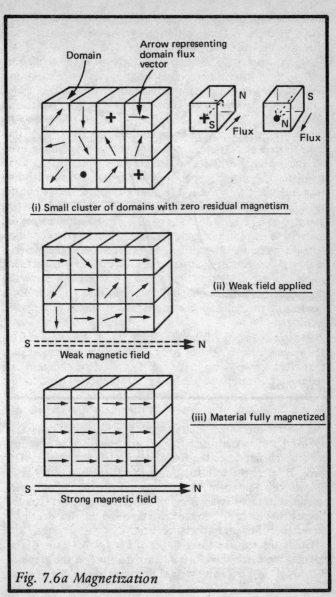

Fig. 7.6a Magnetization

241

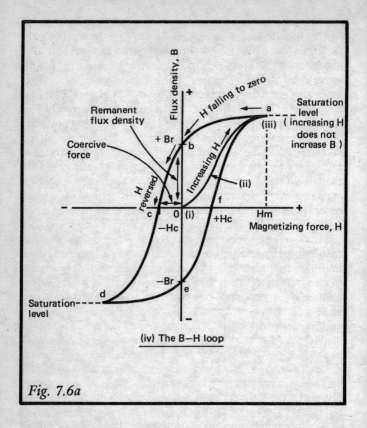

Fig. 7.6a

[or ampere-turns per metre — before SI units, known as *oersteds*]. It is the magnetizing force created in a recording head which impresses the signal magnetically on a tape. The force, H gives rise to a total magnetic flux Φ which has its *flux density* equivalent, B , the amount of flux per unit area, measured in teslas (webers/m²). The relationship between the magnetizing force H applied to a tape and the flux density B it creates is of major importance here.

Going through the three stages of Fig.7.6, in (i) there is no magnetizing force therefore no flux density, on the graph of B and H in Fig.7.6(iv) we are therefore at the origin, point

242

(i). In (ii) H has increased and because the domains are partly aligned there is some flux density and B has a value as shown on the curve at (ii). Further up the curve at (iii) B is maximum and because all domains are aligned any further increase in H is ineffective. These three points only cannot determine the complete curve but a typical one has been sketched in (O–a). Now if H is reduced it is found that although some domains slip back into random positions, others remain aligned even when H is zero, the very effect necessary to retain a signal on tape. This is demonstrated in Fig.7.6(iv) by the section of the curve a–b and at H = 0 there is a flux density remaining, +Br . This is known as the *remanent flux density* and the material is said to possess *remanence* (the term *retentivity* is sometimes used, i.e. the ability to retain magnetism). To restore the sample of material to the condition of (i) in Fig.7.6 (all domains random), H has to be reversed, that is a magnetic field must be applied in the opposite direction to that shown in (ii) and (iii). B then falls as shown by b–c of the curve, eventually to zero and the value of −H required for this is known as the *coercive force*, −Hc (it coerces the domains to give up their alignment). Increasing −H above −Hc begins to align the domains in the opposite direction (B becomes negative), rapidly at c but reaching saturation at d as fewer and fewer domains are left to align. Similar reasoning applies as −H is reduced back to zero (d–e), reversed and then increased in the original direction (e–f–a). The curve a, b . . . f , a is appropriately known as a *B-H loop*, also as a *hysteresis loop* (from Greek, "coming after" − B lags behind H). For optimum recording and retention of a signal on tape therefore both remanence and coercivity of the material should be high.

Note especially from Fig.7.6(iv) that the whole of the B-H loop is curved so we must expect problems of non-linearity.

7.2.2 The Tape

At the heart of the system is the recording medium, the tape. There are several widths available from some 5 cms for certain video recorders down to 3.8 mm in the popular *audio cassette* on which the magnetic track is only 0.6 mm wide. Dimensions are smaller still in *microcassettes*. Such cassettes are now so commonplace that there is little need to explain their operation except perhaps first to get tape size and tracks into perspective. Also because this is a book mainly concerned with electronics we will not consider the mechanics of the tape drive system, we will also use the audio cassette as an example, knowing that the basic electronic principles are the same for all tapes and systems. Fig.7.7 shows the disposition of the four separate tracks which can be accommodated on an audio cassette tape. The standard arrangement is for stereo. Dimensions shown are approximate as they vary according to the manufacturer and the particular coating and base used. Cassette recording and replay heads are therefore double units catering for two separate tracks (L and R channels) at once.

The *tape base* is of pliable yet strong plastic such as cellulose acetate or polyester with the latter perhaps most commonly used. The *tape coating* consists of a magnetic

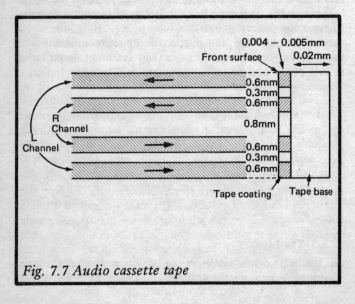

Fig. 7.7 Audio cassette tape

oxide powder mixed with a binder which coats the particles to prevent bunching when magnetized and which presents a smoother surface on the tape to minimize head wear. In the coating are also a plasticizer to keep the binder flexible (not with polyester) and a lubricant, again to minimize wear. The magnetic oxide powder is either an acicular (needle-like) derivative of ferric oxide (chemically Fe_2O_3) or chromium dioxide (CrO_2). Although described as "needle-like", the length of a particle is actually only around three times the width, the length being of the order of 0.5×10^{-3} mm. In manufacture the particles (not the domains) are aligned to a certain extent lengthwise along the tape, this enables more particles to be accommodated than if they were scattered randomly so there is a greater concentration of the oxide. The preceding section shows that high remanence and high coercivity are required, the ranges for audio tape are from about 0.08–0.16 T (tesla, $= Wb/m^2$) remanence and 20,000–40,000 A/m (or At/m) coercivity. The coercivity is of the same order as found in a metal alloy permanent magnet but the remanence is lower, magnets have values up to around 1 T as would be expected from their greater densities compared with an oxide powder.

Materials with high remanence and coercivity and therefore most suitable for permanent magnets are said to be magnetically (not physically) *hard*. We meet the soft material in the next section.

7.2.3 Magnetic Tape Heads

We all must have studied the fundamental principles which govern the operation of tape heads many times before.[1/5] Fig.7.8 shows the elementary features of a single head. It is built on a ring-shaped magnetic core with two gaps in the magnetic circuit, one where it is in contact with the tape, the other opposite at the back. The core is formed from a laminated special metal alloy or ferrite or ferro ceramic material, these in contrast to the oxides used in the tape have both low remanence and coercivity and are known as *magnetically soft* materials (in earlier days all materials with these properties were mechanically soft, i.e. ductile and

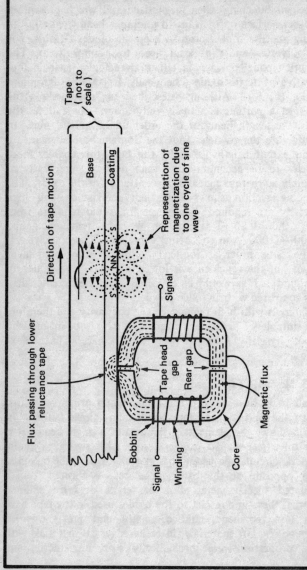

Fig. 7.8 Magnetic tape head

malleable, nowadays this does not necessarily apply). The purpose of the core is to provide a low reluctance path for the magnetic flux set up by the signal current when recording or induced from the tape on replay and because the flux varies at signal frequency it must be capable of changing quickly with the core material retaining as little magnetism as possible. The B-H loop for a magnetically soft material is therefore very narrow and of small area so that Br and Hc are small.

7.2.3.1 Record and Erase Heads

Because the tape coating bridges the front head-gap, flux which would normally be forced to flow through the comparatively high reluctance of the gap, finds an easier path through the much lower reluctance of the oxide particles of the tape coating. The magnetizing force H therefore leaves a remanent flux in the tape according to the B-H loop for the particular magnetic material. The tape flux is expressed in terms of the width of the track in nanowebers per metre (nWb/m = Wb x 10^{-9}/m) as measured by a specified reproducing head and generally tapes work at a few hundred nWb/m. On the recording head a gap at the rear serves to reduce non-linearity caused by changes in core permeability at different levels of flux. It puts a high reluctance in series in the magnetic circuit, partly swamping the lower reluctance of the core and hence its variations. To restore the total flux a larger signal current must be provided than with no rear gap.

As Fig.7.8 shows, the result in magnetic terms for a sine wave can be looked upon as a pair of bar magnets within the tape coating, the magnet length depending not only on the frequency of the wave but also on the speed of the tape. To get this into perspective, cassette audio tape moves at 4.75 cm/s (v) and designating the wavelength of the signal on the tape by λ, since $\lambda = v/f$,

$$\lambda = \frac{4.75 \text{ cms}}{1000} = 0.0475 \text{ mm for } f = 1 \text{ kHz}$$

and one tenth of this when $f = 10$ kHz, or at 1 kHz 210 cycles are crowded into 1 cm of the tape, and at 10 kHz, 2100.

7.2.3.2 Replay Head

Although Fig.7.8 is used mainly to explain the record head (erasure follows later) it can equally demonstrate the principles employed in the reproduction (replay) process. Put simply, magnetization along the tape induces a flux into the head at the point of contact, the flux taking the easier path through the core rather than through the gap and therefore because it is changing inducing an e.m.f. into the windings.[1/5.3]

7.2.4 Magnetization

We became suspicious in Sect.7.2.1 that because of the curved shape of the B-H loop, distortion would arise in the recording process, that is, that the magnetic flux remaining on the tape would not be proportional to input signal level. Fig.7.6(iv) tells only part of the story, simply that a magnetizing force H_m results in a remanent flux density, B_r. For values of H below H_m smaller loops are produced, each resulting in a different value of B_r and a typical complete curve for both directions of H and known as a *tape transfer characteristic* (t.t.c.) is shown in Fig.7.9(i). It is immediately evident from this that if H follows a sine wave as shown, the variation in B_r is certainly not a sine wave. To operate only on the straight section of the t.t.c., a constant value of +H or −H could be used as a bias (d.c. biassing), this however uses only the +ve or −ve half of the characteristic. A more satisfactory method is adopted of *h.f. biassing* as shown graphically in Fig.7.9(ii). A sine wave at 3−4 times the highest signal frequency is the bias and it is mixed linearly with the signal so that the bias frequency amplitude varies as shown. This is *not* modulation of an h.f. signal by an a.f. one[5/5.2] for this requires non-linear mixing. However, unwanted non-linearities create some modulation and this is the reason the bias frequency is several times the maximum audio frequency. This places modulation products outside of the audio range

248

and so prevents them from creating noise. As an example, with a bias frequency f_b of only 25 kHz and audio frequency f_a of 15 kHz, the side frequencies on modulation are $f_b + f_a = 40$ kHz (no problem) and $f_b - f_a = 10$ kHz, an unwanted audio noise signal. Clearly making f_b say 50 kHz or more eliminates this source of interference.

As Fig.7.9(ii) shows, using a bias frequency of the correct amplitude brings the audio signal onto the straight part of the t.t.c. The h.f. bias, being well above audio is of no concern on replay thus the +ve and −ve audio flux variations add together resulting in the overall variation as shown, relatively free from

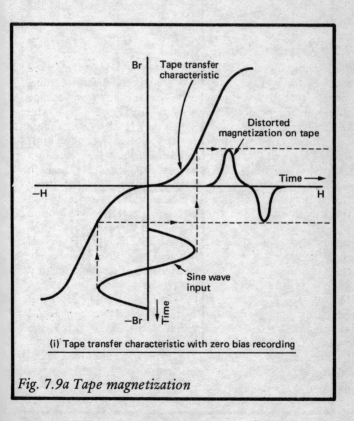

(i) Tape transfer characteristic with zero bias recording

Fig. 7.9a Tape magnetization

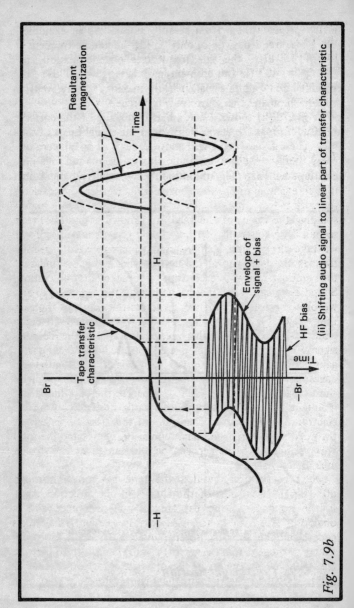

Resultant magnetization

Time

Tape transfer characteristic

H

−H

Br

−Br

Envelope of signal + bias

HF bias

Time

(ii) Shifting audio signal to linear part of transfer characteristic

Fig. 7.9b

250

amplitude distortion. Note that the very non-linear section of the t.t.c. as it passes through the origin is not used and also that an enhanced magnetization is obtained because both the +ve and −ve sections of the t.t.c. contribute. Fig.7.9(ii) reminds us also that the amplitude of the bias is critical so because the t.t.c.'s of, for example, ferric oxide and chromium dioxide coatings are different, a recording machine may need adjustable bias levels to accommodate both types.

7.2.5 Reproduction

The tape head gap (Fig.7.8) in the replay head is all important, it is an example of precision engineering because for a cassette recorder it is no more than a few micrometres. Truly understanding the effect of the head gap (known as *gap loss*) may seem a little elusive at first but should give us no difficulty if we take the process step by step. Imagine we are peering through the head gap at the tape coating in Fig.7.8 and that the tape stands still each time we make an observation. Suppose we can "see" the magnetization through the gap and we plot a graph of the flux ϕ across the width of the gap, then (a) of Fig.7.10(i) is the picture when the effective gap width ℓ is exactly half the wavelength ($\lambda/2$) of the recorded signal. The term "effective" is used because when dealing with gaps a few μm wide, compared with this dimension, the gap edges cannot be perfectly smooth nor can any non-magnetic shim (spacer) or plating used to separate the core limbs and maintain the gap be of uniform width itself. Thus ℓ is electronic rather than mechanical. At this stage ϕ is zero at the edges of the gap and maximum at the centre and the total flux seen through the gap can be represented by the shaded area under the curve.

Next at (b) imagine that the tape has moved through 60° ($\lambda/6$). Now some of the flux is in one direction, some in the opposite so the net flux is the difference and is obviously less than in (a). At (c) the +ve and −ve fluxes cancel and the net flux is zero. After this, as the tape continues to move across the gap (d) and (e) demonstrate that the flux rises to maximum in the opposite direction to that at (a) as shown to the right in the Figure. Readers who have

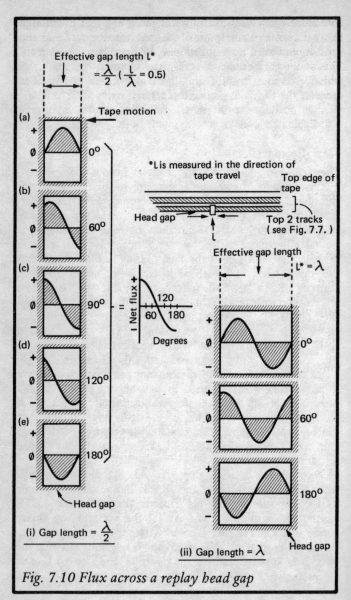

Fig. 7.10 Flux across a replay head gap

252

studied mathematical calculus will recognize that the change in total flux as a sine wave of magnetization moves over the gap is represented by a cosine wave ($\int \sin x \,.\, dx = -\cos x$). This is merely a sine wave displaced by 90° and it is this which gives rise to the e.m.f. generated in the replay head according to the basic principle of generators:

$$e = N \frac{d\phi}{dt}$$

where N = number of winding turns and $d\phi/dt$ represents the *rate* of change of flux within the core. The rate of change of flux is indicated by the *slope* of the net flux curve in the diagram.

Next in Fig.7.10(ii) we see the same frequency but double the gap length, i.e. $\ell = \lambda$ and it is evident that the +ve flux always cancels the −ve resulting in zero output simply because there is always a total of half a cycle above and the same below the axis. The same result would arise if in (i) the frequency were doubled for then again $\ell = \lambda$.

What is clearly having considerable effect is the ratio ℓ/λ for at $\ell/\lambda = 0.5$ there is maximum output, at $\ell/\lambda = 1.0$, there is zero. In fact, working in decibels the formula becomes:

$$\text{output level} = 20 \log \sin \left(\pi \frac{\ell}{\lambda} \right) \text{dB} \text{, with } \pi \text{ in radians, or}$$

$$20 \log \sin \left(180° \frac{\ell}{\lambda} \right) \text{dB,}$$

working directly in degrees. This is easily checked:

at $\ell/\lambda = 1.0$ output $= 20 \log \sin 180° = -\infty$ dB

at $\ell/\lambda = 0.5$ output $= 20 \log \sin 90° = 0$ dB

at $\ell/\lambda = 0.1$ output $= 20 \log \sin 18° = -10.2$ dB

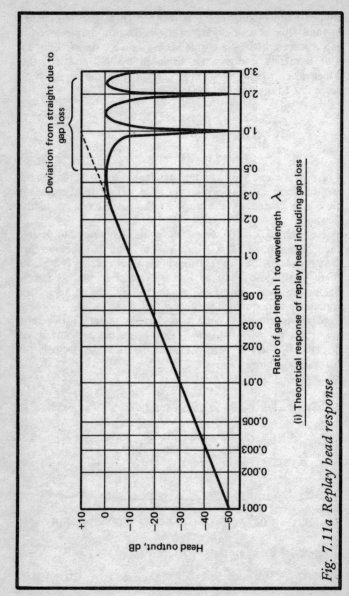

Fig. 7.11a Replay head response

(i) Theoretical response of replay head including gap loss

Ratio of gap length l to wavelength λ

Head output, dB

Deviation from straight due to gap loss

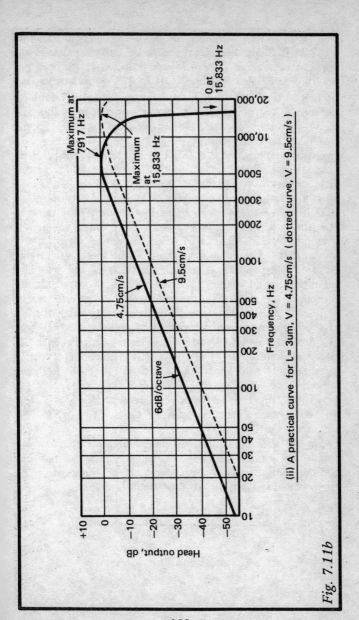

(ii) A practical curve for L = 3μm, V = 4.75cm/s (dotted curve, V = 9.5cm/s)

Fig. 7.11b

255

and so on so that we ultimately build up the graph shown in Fig.7.11(i). This is a general theoretical curve, a practical one for a cassette tape (v = 4.75 cm/s) and effective head gap length 3 μm follows in Fig.7.11(ii). The output curve rises at first at 6 dB/octave (i.e. the head output voltage doubles) because e is proportional to the *rate* of change of flux and if the frequency doubles, the rate doubles. As the gap loss begins to take effect with rising frequency the curve turns over with maxima at $\ell/\lambda = 0.5, 1.5, 2.5$ etc. or in the practical case at about 7,900 Hz and with zero output at double this frequency ($\ell/\lambda = 1.0$), i.e. 15,800 Hz. However, if the tape speed on both record and replay is doubled, λ is doubled so ℓ/λ is halved and we would find that it reaches the value of 0.5 (for maximum output) at twice the previous frequency. This is shown in Fig.7.11(ii) where the dotted curve demonstrates that the effect of doubling the tape speed is to double the frequency of output at each decibel level, the especially important point being that the frequency of maximum output in this particular case makes the system effective to well over 20 kHz (less than 2 dB down at 22 kHz) whereas at the lower tape speed the output has faded out completely at just over 15 kHz. In audio practice the output above the first zero point [$\ell/\lambda = 1$ in Fig.7.11(i)] is unusable for clearly equalization would be difficult. This simple example shows why tape speeds other than 4.75 cm/s are in use, these are mostly for professional work and are 9.5, 19.05, 38.1 and 76.2 cm/s being the metric equivalents of 3.75, 7.5, 15 and 30 inches/s respectively.

We can also see a means of determining the effective gap length, for example by reproducing a test tape which has constant magnetization with frequency and observing the frequency of the first zero in the output, f_0 . Then

$$\lambda = \frac{4.75 \text{ cm/s}}{f_0} \quad \text{and} \quad \ell = \lambda$$

therefore

$$\ell = \frac{4.75 \times 10^4}{f_0} \ \mu m$$

and similarly for any other tape speed.

Besides the gap loss there are several others, two of significance are eddy current losses[2/3.8.4] in the head core and a loss due to the fact that the tape is never in perfect contact with the head.

What we have discovered so far is that even if the recorded tape flux is constant with frequency, the e.m.f. contained from the replay head is certainly not constant, it rises with frequency at the lower end of the audio range and falls rapidly at the higher end depending on the tape speed and head gap. Equalization (Sect.6.1.2.2) in the replay chain takes care of this but there are also difficulties in obtaining a constant recorded tape flux with frequency even though arrangements may be made to maintain the recording head signal current constant. There are several reasons for this, especially imperfect contact between tape and recording head, fall in tape permeability at the higher frequencies and some erasure by the bias at these frequencies (*erasure* follows in the next section). Thus so that a tape recorded on one machine may be replayed correctly on any other both the record magnetization and the replay equalizations are standardized in the form we have found with disk recording and reproduction (Sect. 7.1.4.2), that is, by quoting a series of time constants for the various tape speeds and tape coatings. Because we have already looked at equalization in some detail in Sect.6.1.2 it is not repeated here except to outline briefly the procedure.

The typical curves in Fig.7.12 show that some pre-emphasis is provided in the recording amplifier at the higher frequencies so as to raise these further above the noise to improve the signal/noise ratio on the tape. This results in the requirement of additional replay equalization, the total used being typically as shown. This is comparable with but not directly applicable to Fig.7.11(ii) because the latter

257

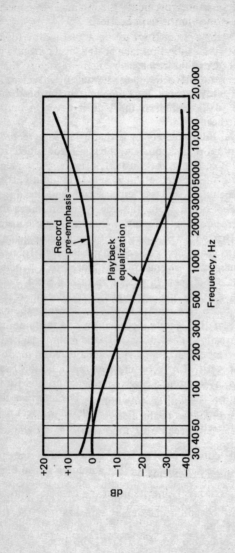

Fig. 7.12 Typical recording and replay characteristics

excludes the effects of losses other than that of the gap. Further signal/noise improvement is obtained by special techniques as described in the next Section.

In practice a tape system is set up by adjusting the replay equalization so that a calibration tape gives a level output with frequency. The record pre-emphasis is then adjusted so that a recording of a constant level input signal at all frequencies also gives a level output on replay. The tape magnetization is then the same as that of the calibration tape.

7.2.6 Erasure

The restoration of the magnetic domains to their random magnetizations after alignment in the recording process is known as *erasure*. The tape coating is subjected to several cycles of an alternating magnetic field of sufficient magnitude for saturation [Fig.7.6(iv)] impressed across an erase head gap. As the tape moves away from the gap the strength of the field it experiences diminishes, in effect the domains are subjected to cycles of magnetization which can be described by ever decreasing B-H loops. Ultimately the loop is so small that the values of H become insufficient to reverse the magnetization of the domains and they cease to be affected, this occurs in a random manner. So that the tape crossing the erase head gap experiences several cycles of saturation, the erase head gap is comparatively large, of the order of 20 μm. The head therefore needs no rear gap because high series reluctance is now provided by the front gap. A suitable erase frequency is that of the bias, say above about 50 kHz, in many systems one oscillator only is required to supply both recording bias and erase. The current fed to the erase head is naturally higher than that for biassing because of the need to swing the tape well into saturation.

7.2.7 Noise Reduction Systems

Noise in the form of hiss is inevitable in the basic tape process because of the random distribution of tape particles, meaning that equal areas of tape do not contain equal numbers of particles. The record/replay system itself also introduces hum and crosstalk from adjacent tracks, also low level components from incomplete erasure. Generally the noise level is constant with frequency and is only objectionable on very quiet signal passages when it can be as loud as the signal itself. In a given system the signal is recorded at as high a level as possible consistent with avoidance of distortion due to saturation of the oxide [Fig.7.9(ii)], further increases to improve the signal/noise ratio are therefore inadmissible.

7.2.7.1 *Compandors*

Compressor and expander systems (*compandors*) used over radio channels to improve the signal/noise ratio are well known. In these the high-level signals are first reduced (compressed) and the whole signal train then amplified so that effectively overall the low-level signals receive greater amplification. Thus after transmission the low-level signals appear at the receiving end at an enhanced level above the noise and as shown in Fig.7.13, when expansion is applied this improved signal/noise ratio is maintained. Such systems are used successfully in magnetic tape recording where the record/replay machine replaces the radio channel.

7.2.7.2 *Dolby B*

A system which has some similarities with the compandor but processes only the lower level components of the signal and is in common use in domestic magnetic recorders is the Dolby B (developed by Dr Ray M. Dolby, a British engineer). The Dolby A system is more complicated and is used in professional work.

The fundamental frequencies of music and speech are in a range up to about 4,000 Hz, harmonics continue higher. Fourier analysis[(2/1.4)] shows that for repetitive waveforms the harmonics decrease in amplitude compared with the fundamental as frequency rises thus generally the energy content of a signal appearing at the input of a recording system is greatest at the lower frequencies, falling steadily as frequency rises.

A little digression will not come amiss here for us to fully understand energy (or intensity) *spectrum* diagrams. They are merely a series of upright lines on a frequency basis, each line representing the energy or intensity theoretically at its particu-

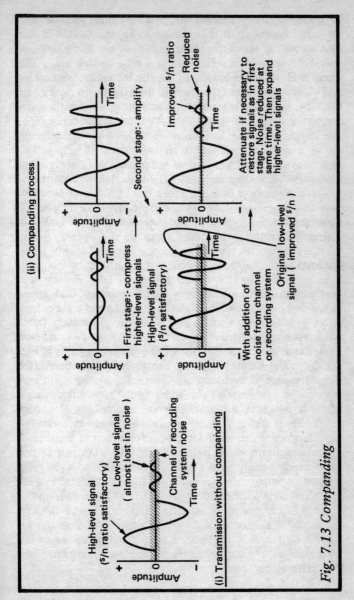

(ii) Companding process

Amplitude

First stage:- compress higher-level signals

High-level signal (s/n satisfactory)

Time

Second stage:- amplify

Time

Amplitude

With addition of noise from channel or recording system

Time

Original low-level signal (improved s/n)

Amplitude

Improved s/n ratio

Reduced noise

Time

Attenuate if necessary to restore signals as in first stage. Noise reduced at same time. Then expand higher-level signals

Amplitude

High-level signal (s/n ratio satisfactory)

Low-level signal (almost lost in noise)

Channel or recording system noise

Time

(i) Transmission without companding

Fig. 7.13 Companding

lar frequency but practically within a very small band centred on that frequency. It is important to appreciate that in this case we are not measuring the amplitude of the whole signal as would be seen on a meter but in a way looking inside the signal itself at some instant to see the individual frequencies and their amplitudes (spectrum is from Latin, to look). In Fig.7.14, (a) is the spectrum of a sine wave, it comprises one frequency only, (b) shows that for a particular organ pipe, the sound is rich in harmonics yet (c) for a different pipe shows a note lacking almost all harmonics. The practical variations are large but this does not prevent us from choosing a representative diagram as in (d) for explanation purposes.

On a frequency basis therefore the higher level components of a signal are those of low frequency and they suffer little from added system noise. Accordingly the Dolby B system looks at the incoming signal on a frequency basis, takes no action on the low frequency components but raises the level of the higher ones, reducing the lift as the incoming level increases. In this way as we will see later, distortion is kept to a minimum. Effectively high frequencies which arrive at low level are carried through the recording and replay chain at an enhanced level and subsequent complementary treatment restores the signal.

A simplified block diagram of the process is given in Fig. 7.15(i). The loops containing the variable high pass filter (h.p.f.) system at each end are identical. At the input, low frequencies pass through the adder to the record/replay unit while higher ones for processing are selected by the fixed h.p.f. These higher frequencies are then passed to a variable h.p.f. system which raises its cut-off frequency as signal level increases thus rejecting frequencies at the lower end of the band which already have a sufficiently high signal/noise ratio to need no amplification, accordingly these also take the direct route and pass via the adder unaffected. The output from the variable h.p.f. system is amplified and mixed with the low frequency signal components in the adder. After replay the second variable h.p.f. system removes the high frequency lift by means of a negative feedback loop to a subtracter and thus restores the signal.

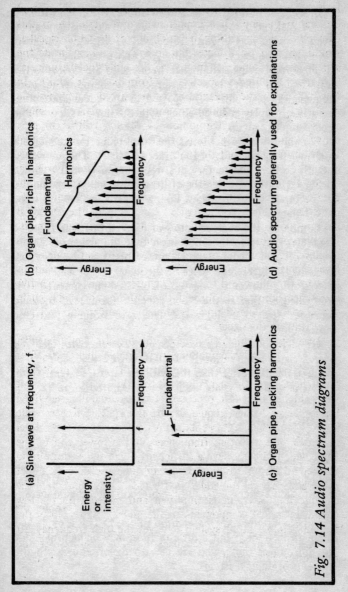

(a) Sine wave at frequency, f

(b) Organ pipe, rich in harmonics

(c) Organ pipe, lacking harmonics

(d) Audio spectrum generally used for explanations

Fig. 7.14 Audio spectrum diagrams

263

Bearing in mind that, as shown above, the level within a given signal gets less as frequency rises, the explanation can be given pictorially as in Fig.7.15(ii) where each diagram shows the signal frequency spectrum. At (a) we see that with no "Dolby" processing, the higher frequencies are well into the noise and (b) to (e) how with processing the upper range of frequencies is amplified and passed to the adder and subsequently on replay reduced along with the noise added by the record/replay system.

For the lowest signal levels the noise reduction at medium and high frequencies is of the order of 10 dB. The pre-record h.f. lift and complementary post replay reductions are shown in Fig.7.15(iii). The degrees of lift and reduction become less as the input signal increases, falling to zero for a signal some 40 dB higher.

Comparing the characteristics of Fig.7.15(iii) with those of Fig.6.8(i) may give the impression that the system is simply treble lift followed by treble cut. This is only so for low levels of the high frequencies of the signal. What is important is that in addition the *degree* of lift and subsequent cut is controlled so that as these components rise in level they are processed less so avoiding the danger of running into tape saturation.

Given a good quality record/replay system, with Dolby B overall signal/noise ratios of 60 dB or more are possible and as much as 10 dB better than this with the later C system. Complete Dolby B circuits are available in integrated form.

7.3 DIGITAL RECORDING

Although not in common use at the present time, it is worthwhile looking briefly at the techniques and prospects of digital recording based on the knowledge of the success of digital transmission systems. To fully appreciate what digital recording has to offer we must first know something about the principles of *pulse-code modulation* (p.c.m.). This is explained at length in Book 5 but a little revision may be helpful and it also introduces those who are new to p.c.m. at least to the elements of digital systems.

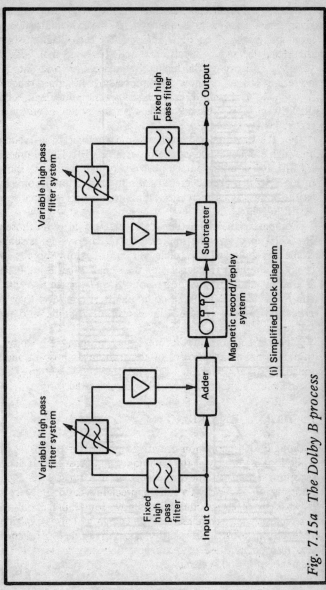

(i) Simplified block diagram

Fig. 7.15a The Dolby B process

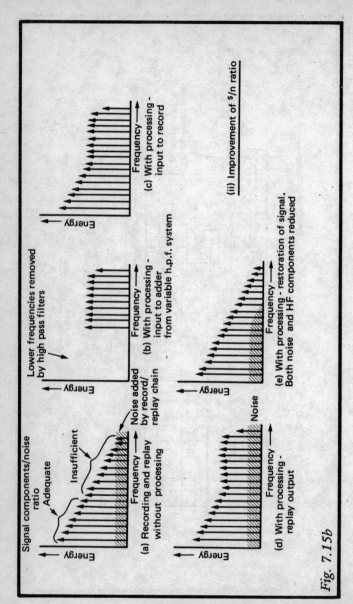

Fig. 7.15b

(a) Recording and replay without processing

Signal components/noise ratio

Adequate

Insufficient

Noise added by record/replay chain

Energy → Frequency →

(b) With processing - input to adder from variable h.p.f. system

Lower frequencies removed by high pass filters

Energy → Frequency →

(c) With processing - input to record

Energy → Frequency →

(d) With processing - replay output

Noise

Energy → Frequency →

(e) With processing - restoration of signal. Both noise and HF components reduced

Energy → Frequency →

(ii) Improvement of s/n ratio

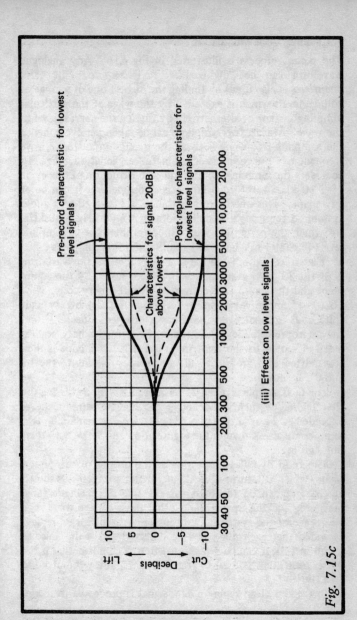

(iii) Effects on low level signals

Fig. 7.15c

7.3.1 Pulse-Code Modulation

The p.c.m. process is illustrated in Fig.7.16. Any analogue waveform can be "digitized". It is *sampled* (the term *quantized* is also used) by finding the highest one of a range of voltage levels which is exceeded by the wave at the particular sampling instant or alternatively by finding the nearest level to the wave. At least two samples must be taken per cycle, hence the sampling frequency is twice the maximum analogue wave frequency. For a more meaningful explanation, Fig.7.16 considers the sampling, coding and reconstruction of one cycle of a 10 kHz waveform, shown as sinusoidal but it can be of any shape. This one is from within a 20 Hz − 20 kHz audio system so requiring a sampling rate at 2 x 20 kHz = 40 kHz, for which the period is 25 μs. Suppose there are 16 sampling levels (e.g. at 1 mV intervals) and that the first sample is taken at t = 15 μs. From (i) in the Figure we see that the waveform is at the 13 (mv) sampling level. This is noted by the equipment and the figure 13 is converted into its binary code. Now 16 (0−15 inclusive) sampling levels require a 4 binary digit (bit) code so that every level has a different one[4/A1] and the appropriate code for 13 using the normal binary equivalents as shown to the right in (i) is 1101. This code is then transmitted as a group of pulses or spaces relating to the 1's and 0's as shown.

The next sample is taken 25 μs later at t = 40 μs. Using the first technique mentioned above the highest sampling level exceeded is 11 resulting in a code 1011. The remaining two samples are also shown in the Figure at 65 and 90 μs with their pulse groups.

The train of pulses appearing at the output of this equipment (which is a form of analogue/digital converter) is shown at (ii) and can be recorded for example on magnetic tape, inevitably picking up tape and other noise on the way. The advantage of the system now becomes apparent for on replay, provided that each pulse can be detected above the noise as it can in (iii), it can be *regenerated* free of all noise, that is the pulse with noise is used to trigger a new pulse without noise [see (iv)].

In (v) the clean replay p.c.m. signal is processed to recon-

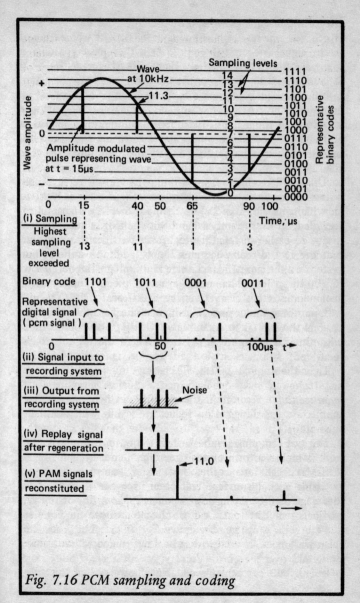

Fig. 7.16 PCM sampling and coding

269

stitute the original pulse amplitude modulation (p.a.m.) signal which was generated on sampling as in (i). The whole train of p.a.m. signals occurring every 25 μs is then passed through a low pass filter to regain the original 10 kHz wave.(5/5.5.3.3) It is now possible to appreciate the difference between p.a.m. and p.c.m. signals. The p.a.m. pulse as in (i) and (v) has an amplitude directly representing the voltage of the analogue waveform at the instant of sampling. On the other hand as in (iii) and (iv), all p.c.m. pulses are of the same amplitude but several of them are required to carry the same information as the single p.a.m. pulse.

Using the technique therefore, when after replay, new pulses are regenerated the recording system noise is lost and this happens no matter how many times the signal is re-recorded. On the other hand with analogue recording as discussed earlier in this Chapter, passing a signal through two or more record/replay systems simply adds the noise of one system to the next, in fact there is a limit to the number of re-recordings possible, not so using p.c.m., noise never accumulates, it is all lost at each regeneration.

Everything looks perfect until one checks the level of the original wave in (i) for example at $t = 40\,\mu$s and compares it with the p.a.m. signal in (v) from which the wave is to be recovered. In our example the latter has a magnitude of 11.0 whereas the original is 11.3. There is a discrepancy, known as *quantizing error*(5/5.5.3.1) which actually gives rise to *quantization noise* within the system. So here we are now with a system designed to reduce noise but which actually generates its own! However it is obvious that the greater the number of sampling levels the less is the magnitude of each error and in fact a p.c.m. high quality audio system (up to 20 kHz) might use over 65,000(4/A1.4) sampling levels, so requiring a 16 bit code to represent each one [a code of 16 bits (1's and 0's) can accommodate $2^{16} = 65,536$ binary numbers, all different]. For obvious reasons in Fig.7.16 a 4-bit code only can be shown. With a 16-bit code the quantization noise is extremely low compared with tape noise and over a recording/replay system a signal/noise ratio of over 90 dB can be achieved. Note however that this noise is not lost on regeneration, it has become part of the signal.

7.3.2 Magnetic Tape Systems

Some of the advantages of p.c.m. in magnetic tape systems are evident from the section above. That digital recording on magnetic tape or disk is feasible is shown by its use with computers although the need in this case is for large storage capacity, not particularly for noise reduction. For the latter advantage, as one might expect, there is a price to pay and this is mainly in the bandwidth required. For analogue audio recording the frequency range is simply up to 20 kHz or less and most of us know from experience that excellent results are obtained from cassette tapes moving at a mere 4.75 cm/s. But now we are suggesting recording up to 16 bits every 25 μs which is equivalent to a bit rate of over 600,000 bits/s. Considering that a bandwidth greater than this number is required to accommodate at least the lower harmonics of the pulse spectrum [illustrated in Appendix 2, Fig.A2.2(v)] it is a frightening requirement and more so if we think about 600,000 pulses recorded over a mere 4.75 cm of tape. However, there are special sampling techniques which reduce the need considerably and most music recording will not have such stringent requirements but it remains obvious that the record/replay system must be capable of operating over a large bandwidth, that is, very high tape speeds are required. Hence although *video* recorders demonstrate that ample bandwidth can be obtained from a magnetic tape system, at the present time digital recording is likely to remain solely in the hands of the professionals.

7.3.3 Memory Systems

Integrated circuits to carry out the process of digital recording are plentiful and cheap, thus if a static digital memory[4/4.5] were provided instead of magnetic tape to hold the p.c.m. signals the comparatively high cost of tape transport mechanism, cost of head wear and mechanical equipment maintenance is avoided. But we now know that for *ideal* audio operation the memory required must hold 16 bits for each 25 μs, that is, 40,000 16-bit words for only one second recording, half this for something a little less than perfect and down to 8,000 8-bit words for commercial speech, even then a formidable requirement but obviously less so in the not too distant future.

CHAPTER 8. MAKING MUSIC

There's sure no passion in the human soul,
But finds its food in music.

George Lillo

Above is a reflection as true today as it was when written at the turn of the 18th Century. But now the musical scene is expanding and the expression "making music" no longer relates solely to a band of musicians with their instruments but in addition to perhaps a single artist with one keyboard, a tape recorder and many knobs to twiddle. Electronics has now made the generation of any musical note, be it a faithful copy of that of an existing instrument or one hitherto unknown, simplicity itself through the *music synthesizer*. It is the working of this box of tricks which occupies us here and whether we love, hate or simply tolerate the sounds it makes has no bearing whatsoever on the discussion.

Music, the dictionary tells us, is "the art of combining sounds with a view to beauty of form and expression of emotion". This is undoubtedly the province of the musician, ours on the other hand is the creation of the basic sounds he or she puts together.

The technicalities of conventional musical instruments have already been considered in Sect.3.2 and the following short discussion about music itself will be found useful in understanding how it all fits together.

8.1 MUSICAL SOUNDS

It began a long time ago. Even the great Pythagoras himself took time off from his right-angled triangles to study the effect of plucking pairs of strings for sounds that were pleasing to the ear. Two thousand years later his findings are still written about. Accordingly we are starting this Section with his conclusion, that a musical chord (a combination of two or more single notes) sounds pleasant if the lengths of the vibrating strings are in the ratio of two small integers (whole numbers). The early Greeks did not know *why* one sound was melodious yet another not, today we know no more, therefore we have to accept the definition that a musical chord is almost without exception one which is to human liking, agreeable, harmonious or in the musical term, *consonant*. The less pleasant sounds are accordingly classed as *dissonant*.

8.1.1 Musical Chords

Expanding on this theme, it was found that, considering two notes only, if they have frequencies one octave apart a harmonious sound results. The ratio is 2:1 and both numbers are small integers. Other commonly used ratios are 3:2 (the musical *fifth*, i.e. a span of 5 white keys on the piano keyboard – see Fig.8.1), 4:3 (*fourth*), 5:4 (*major third*), 6:5 (*minor third*), 5:3 and 8:5 (*major and minor sixths*). All of these are harmonious as Pythagoras predicted and as the numbers in the ratios get higher, so dissonance tends to increase. The difference between *major* and *minor* chords is difficult to explain, it can only really be appreciated by listening carefully to them being played, for example (Fig. 8.1) C and E keyed together produce a major third, E and G give the minor third. Major and minor sixths are C and A, E and the C above.

8.1.2 The Musical Scale

The early *Major Diatonic Scale* had 8 notes (the octave) labelled CDEFGABC$_5$ as in Fig.8.1, so spread on musical instruments as to be within the compass of stretched fingers and with frequency ratios (intervals) arranged so that all the chords mentioned above could be played accurately. Fig.8.1 shows the middle octave of the 7 octaves of the piano, each octave is usually distinguished by a number or subscript number attached to the note letter, for example, note G in the lowest octave is G1 or G_1 (we use the latter), in the middle octave G_4, in the top octave G_7. However in Fig.8.1 for the middle octave only, although all notes should have the subscript 4, it has been omitted solely for convenience.

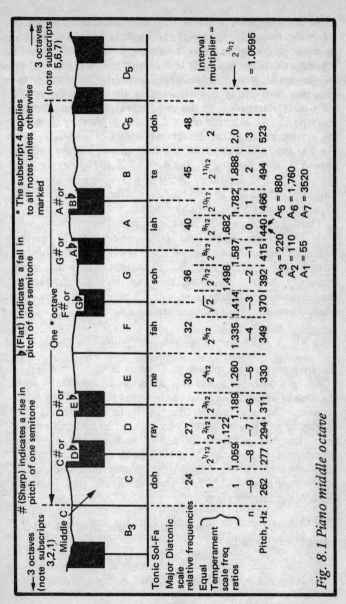

Fig. 8.1 *Piano middle octave*

275

As an example of a chord, the two notes C and F together are in the ratio 32:24, i.e. 4:3 and together sound a "fourth". As music developed however certain imperfections of the scale became evident especially in that change of key could lead to an unwanted switch from major to minor or vice versa. Changing key simply means changing the note from which the octave starts (doh in the tonic sol-fa), it need not necessarily be C. An example of the difficulty is given by changing the key from C to E (i.e. the octave becomes $EFGABC_5D_5E_5$). For the key of C the *major* third (C and E) has the correct ratio 5:4, on changing to the key of E the corresponding third is E and G, a ratio of 36:30 or 6:5, now a *minor* third. The major third has in fact shifted up the octave where the required ratio of 5:4 is given by G and B.

Thus the compromise *Scale of Equal Temperament* was born. Notes were added (the black ones on Fig.8.1) and the difference between any two adjacent notes called a *semitone* (literally half a tone) such that all semitones have the same frequency ratio. Retaining the fact that the frequency doubles over one octave (now with 12 intervals), if x is the interval between two successive semitones,

$$x^{12} = 2 \text{ or } x = \sqrt[12]{2}\,(2^{1/12}) = 1.0595 \, ,$$

the frequency ratios relative to C being as shown in Fig.8.1.

All very well, but surely this no longer agrees exactly with the idea of consonance arising only from ratios involving small integers. This is in fact so, thus for the facility of being able to play in any key we have to accept that all intervals except the octave are not quite in perfect harmony, i.e. they are slightly out of tune. We have become so accustomed to this scale however that the inconsistency passes unnoticed except by those of us who have been given musicians' ears. A single example for illustration is:

major diatonic scale ratio 5:4 = 1.25

major third (C and E) /

   \ equal temperament scale ratio
   $2^{4/12} = 1.26$

a frequency difference of less than 1%.

Internationally a standard musical pitch has been agreed which fixes the note A at 440 Hz. This is shown on Fig.8.1 together with the pitches of all notes in the middle octave (to the nearest whole number) and those of A in the octaves above and below.

8.1.3 Composition

The preceding section gives some insight into musicians' terms and problems, we now look a little more technically into the musical note or chord itself.

A musical sound can be considered as having three main characteristics, *pitch*, which is closely allied to frequency, *timbre* (tonal quality) and *loudness*. Pitch and timbre we have already met on the frequency spectrum diagrams of Fig.7.14, there shown as fundamental plus harmonics. In (a) is a wave with pitch f but no harmonics and therefore no tonal quality, (b) and (c) show notes from two different organ pipes. Even if their fundamental frequencies were the same these notes would sound quite different because (b) has a greater harmonic content than (c), it has a different timbre, not necessarily a more pleasant one. It is mainly the harmonic content which distinguishes each type of instrument.

Loudness must be considered over the short term of the note or chord itself because during this time it is likely to vary greatly. There is of course a general level of loudness of a passage of music but it is the rapid changes in amplitude when a note is generated which give it added character (sometimes called the *expression*) these changes are subdivided generally into *attack* and *decay*. The piano and organ illustrate this well as in Fig.8.2 which demonstrates especially the high rate of attack for the piano but low rate for the organ. For simplicity in the drawing a pure waveform is shown, in practice not only would there be hundreds or thousands of cycles within the two minutes shown but also each wave would be most complex especially since in the case of the piano, more than one string is involved. All other instruments possess their own individual rates of attack and decay.

277

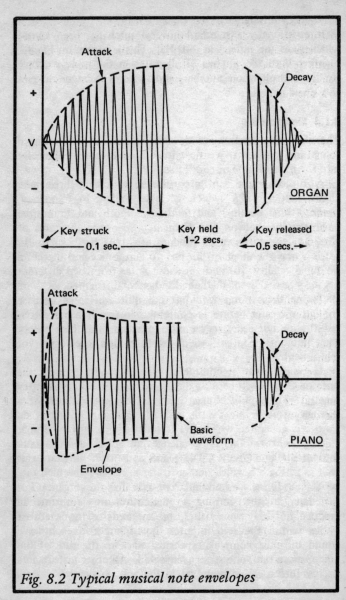

Fig. 8.2 Typical musical note envelopes

Peculiarly enough, *noise* is also a feature of some musical instruments, it is deliberately introduced by the "snare" wires of a side drum and occurs naturally with most wind instruments (especially the flute) for they generate sound with a background hiss of air. Even string instruments may give rise to a small amount.

8.1.4 Modulations

There are several ingenious devices available to the musician for adding interest to music, three commonly used ones are:

vibrato – this is a rhythmic variation above and below the normal pitch

tremolo – a rhythmic variation above and below the normal amplitude

glissando – a "sliding" in pitch as with the Hawaiian guitar or when the fingers slide up or down a vibrating violin string.

The survey of musical sounds in this section serves as a background as to what is required of electronic systems if they are to copy faithfully conventional musical instruments. That they can do so successfully is in no doubt, moreover we will see in the next section that an electronic synthesizer by the very nature of its ability to vary all waveform and wave envelope parameters at will is also capable of generating musical sounds which have no family ties with any other instrument.

8.2 MUSICAL SOUND SYNTHESIS

At this stage we cannot fail to have realized that all musical instruments set something into vibration at a fundamental frequency with the musical quality enriched by Nature's choice of harmonics. This is the basic ingredient of a musical sound and therefore where we must start with synthesis, that is by basing our system on a variable frequency oscillator rich in harmonics. This is shown in Fig.8.3. It is labelled as a

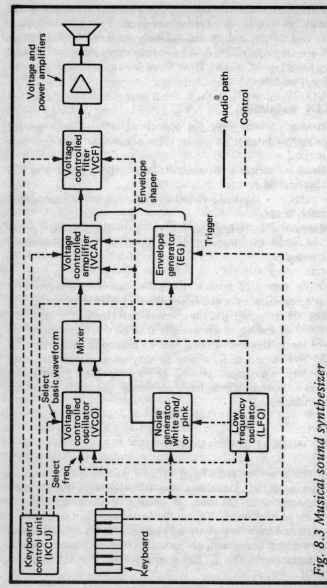

Fig. 8.3 Musical sound synthesizer

voltage controlled oscillator (VCO) because its design is such that the magnitude of a voltage applied from the keyboard to one of the control terminals determines the fundamental frequency of output while a second voltage from the *keyboard control unit* (KCU — a set of keys or switches operated by the user) selects the basic harmonic content. We consider the VCO waveforms in greater detail later but at this stage might refer back to Fig.7.14 (a) and (d) for two particular output spectra occurring for the appropriate control inputs. The keyboard is usually of piano type. Given only this equipment, musical sounds at all audio frequencies can be produced but the tonal range is rather limited and attack, decay and noise (Sect.8.1.3) have not been catered for.

Noise, if required, is added to the output of the VCO in a mixer stage, the output of which passes through a *voltage controlled amplifier* (VCA). Voltage control is supplied from an *Envelope Generator* (EG) which in turn receives its instructions from the KCU and is triggered into action each time a keyboard key is depressed. The controlled variations in the gain of the VCA affect the rate of rise, duration and rate of fall of the signal thus providing the attack and decay features. Together the VCA and EG may be called an *envelope shaper*. A second control input to the VCA from the KCU or a separate pedal adjusts overall loudness of the final output.

The audio waveform now mixed with white or pink noise (Sect.3.4.2) if required and shaped passes to a *voltage controlled filter* (VCF). In the less complicated filters which are low pass only, the input control voltage simply adjusts the cut-off frequency thus frequencies above this value are stopped so reducing the brilliance of the tone if and as required. From the VCF the waveform is voltage and then power amplified as necessary to feed the loudspeaker system.

For vibrato, tremolo and other effects a *low frequency oscillator* (LFO) is used. It has a frequency range of some 0.2 to 20 Hz (vibrato needs up to about 7 Hz) which is variable under control from the KCU and serves to amplitude or frequency modulate the output waveform parameters at the chosen frequency. This is effected at the VCO, VCA or VCF.

This is not any particular system nor in fact can it be classed as a design par excellence, it is simply one of our own, with no frills, arranged for ease of understanding. Most practical arrangements have many additional facilities (e.g. more than one VCO) but certainly the wiring of one unit to another would not be fixed as shown in the Figure, most units would be interconnected via rotary switches, keys or *patching* plugs and cords.

There is little gain from going into detail of music synthesis circuits, there are too many variations. Furthermore IC's have completely taken over even to the extent that single chips are available which contain all the units of a complete synthesizer. Nevertheless, some brief discussion of the main oscillator and what the various waveforms used have to offer is helpful. Also so that perhaps Fig.8.3 may have more meaning, we should see how a control voltage may be set up from a keyboard. This constitutes the remainder of the chapter.

8.2.1 The Basic Waveforms

Of inestimable help in getting to grips with the fundamentals of musical note generation is the method of analysing a continuous periodic waveform (i.e. each cycle repeats regularly) developed by Fourier.[2/1.4] His mathematical proofs are usually the signal for an undignified retreat but the final conclusions are not so intimidating so with the mathematics we have at our disposal we should be able to cope with them successfully. Because of the usefulness of *Fourier analysis* in so many walks of electronic life, examination of several waves (especially those which are most useful in music synthesis) is contained in Appendix 2. Important here are the square, triangular, rectangular and sawtooth forms all of which are pictured in Fig.A2.2. From what the Appendix tells us there is no doubt that these waves together provide a wide range of basic musical sounds. *Looking* at frequency spectra of musical sounds may seem illogical when what we should be doing is *listening* to them. Undoubtedly this would be of interest, especially to the musically minded but we must not forget that we are studying the electronic principles of generating such sounds, not what they actually sound like.

8.2.2 Waveform Generation

The fact that discrete components have given way to integrated circuits in this sphere should not isolate us completely from the design process. From the preceding section and Appendix 2 we understand which waveforms are required, hence the remainder of this Chapter concentrates on the elements of such waveform generation for the oscillator system is the prime constituent of all synthesizers.

8.2.2.1 Tone Generation by Division

Although Fig.8.3 shows a *voltage controlled* oscillator, there is an alternative method of deriving a range of discrete frequencies which has its place in music. Whereas voltage control changes the frequency of the main audio oscillator, this method divides a single fixed high frequency down to the many required. It starts with a *top octave generator* which produces a full octave (12 frequencies) at the top of the musical scale from a single high-frequency input. This is then followed by $12 \div 2,4,8$ etc. frequency dividers for the lower octaves. A very wide range is available by this method so it is ideal for electronic organs and pianos. The generator output is usually in pulse form [rectangular wave, Fig. A2.2(v)] which drives other units or itself is modified as required. Fig.8.4 shows a possible arrangement with a 1 MHz main oscillator divided as shown and connected to a few keys only but sufficient to explain the method. The system requires one top octave generator IC and 12 frequency divider IC's but in fact the whole arrangement is also available on one IC. Using this method all tones are held in strict frequency relationship, therefore single notes cannot go out of tune, while simply altering the input supply frequency tunes the whole keyboard. In practice, to minimize keyboard wiring, matrix arrangements(4/4.5.1) are possible, bringing the keyboard wires down from over 80 as is required in Fig.8.4 to about 20 for a 7-octave spread.

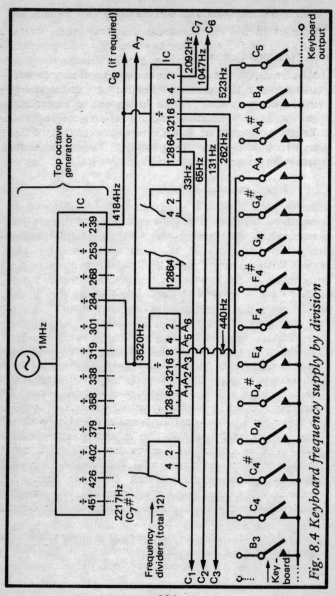

Fig. 8.4 Keyboard frequency supply by division

8.2.2.2 The Voltage Controlled Oscillator

Inherent in the list of musical scale frequencies shown in Fig.8.1 is the fact that adjacent notes do not have a common frequency *difference* but have a common frequency *ratio*. More precisely stated, they are in geometrical progression, not arithmetical. Let us start by making the interval multiplier the constant k $(= 2^{1/12})$, then shifting one note up multiplies the frequency by k , two notes by k^2 , n notes by k^n etc. To put k into a formula we need somewhere from which to start so why not note A at 440 Hz? Then with A as the reference, the frequency of any other note is:

$$f = 440 \times k^n$$

where n is the number of the note counting from the reference (e.g. as shown in Fig.8.1 for A = 0), for example:

C has $f_C = 440 \times k^{-9} = 262$ Hz

E has $f_E = 440 \times k^{-5} = 330$ Hz

C' has $f_{C'} = 440 \times k^3 = 523$ Hz .

The point of this exercise is to show that the frequency scale is *exponential* (n is the *exponent*). Unfortunately this does not match up with the keyboard, which to avoid complication prefers to have a linear control voltage output. Fig. 8.5 clarifies this by showing as an example a simplified arrangement of a 4-octave keyboard and a little work with Ohm's Law (made easy because in this example preferred value resistors are not used) shows that there is 1 V control voltage difference per octave or $\frac{1}{12}$ V difference per semitone (values which are in common use). Obviously therefore the linear control voltage from the keyboard has to be changed to exponential to drive musical note based units such as the main waveform generators and filters.

In most cases the collector current of a transistor varies exponentially as the input base-emitter voltage[3/2.2.4] so giving a linear to exponential conversion. Thus connecting a

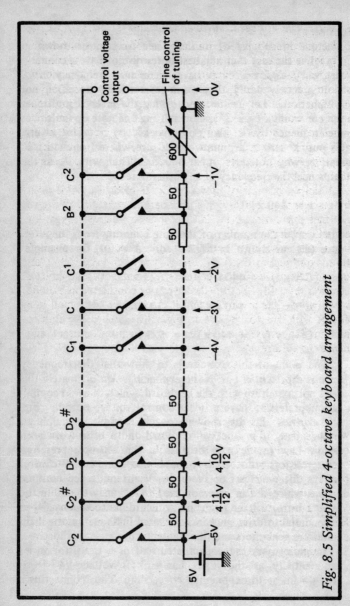

Fig. 8.5 Simplified 4-octave keyboard arrangement

transistor to the keyboard output results in an exponential collector current instead of a linear keyboard control voltage. In practice because even the best of transistors is temperature-sensitive, additional circuitry is necessary for frequency stability and typically an operational amplifier (see Chapter 6) with non-linear feedback to give the overall exponential response would be used, as shown in Fig.8.6. The *exponential generator* would most likely be preceded by an *adder*, probably using the high input impedance of an operational amplifier to add other control input voltages in addition to that from the keyboard.

A simple example of how a VCO can be developed is given by a single stage employing a *unijunction transistor*. We have not met this device so far in this series so we must digress for a brief explanation of its operation.   A unijunction transistor is one with a second base instead of a collector, its construction is outlined in Fig.8.7(i). An n-type silicon slice[3/1.2] has one emitter and two base contacts b_1 and b_2, the emitter is nearer b_2. In effect the device can be considered as a diode with two outlets in parallel and it is normally connected in a circuit with +ve to b_2 as in Fig.8.7(ii). When the emitter junction is reverse biassed there is no emitter current and hence the voltage V_x is

$$\frac{R_{b_1}}{R_{b_1} + R_{b_2}} \times V, \quad \text{(1/3.4.4)}$$

which is +ve. For the emitter junction to remain reverse biassed, V_e must be less than V_x. When however V_e exceeds V_x the junction is forward biassed and current flows between the emitter and base b_1 so creating the normal diode substantial fall in resistance,[3/1.5.2] thereby reducing V_x and increasing the degree by which it is exceeded by V_e. Hence the emitter current increases still further, the effect is cumulative and the transistor suddenly becomes conductive. The triggering point is therefore at

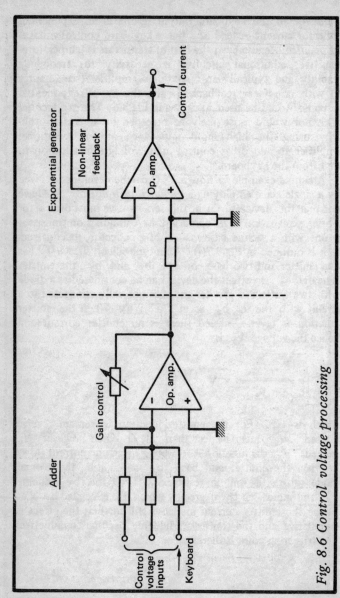

Fig. 8.6 Control voltage processing

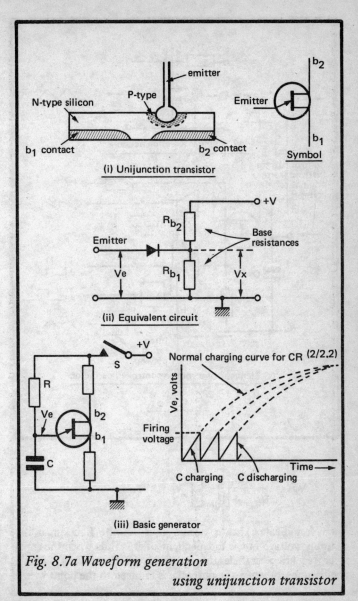

Fig. 8.7a Waveform generation
using unijunction transistor

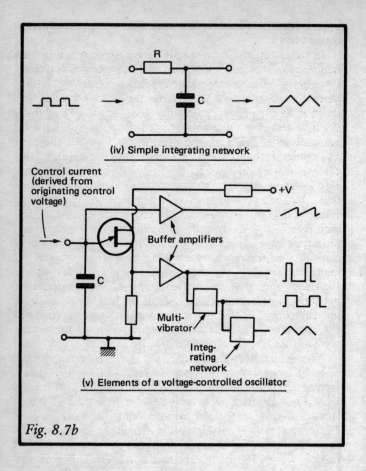

R

C

(iv) Simple integrating network

Control current
(derived from
originating control
voltage)

+V

Buffer amplifiers

Multi-
vibrator

Integ-
rating
network

C

(v) Elements of a voltage-controlled oscillator

Fig. 8.7b

$$V_x = \frac{R_{b_1}}{R_{b_1} + R_{b_2}} \times V$$

which indicates that it occurs at a definite fraction of the
supply voltage but is independent of it. Thus when the emitter reaches a sufficiently +ve potential the devices "fires",
subsequently when the potential is reduced to the point where

V_e is less +ve than V_x , it reverts to the high resistance condition.

In use as an oscillator an RC combination controls the frequency as shown in Fig.8.7(iii). When switch S is closed, C commences to charge with the emitter voltage V_e rising positively. Eventually V_e reaches the firing potential and the transistor conducts, so rapidly draining the charge from C thus reducing V_e until the transistor changes back to non-conducting. The graph in (iii) of the Figure shows how V_e varies with time and provided that the values of the components are arranged so that only the lower and reasonably linear part of the charging curve is used, it can be seen that V_e is a waveform of sawtooth shape. Moreover the voltage on b_1 is of a pulse pattern, a single pulse arising each time C is discharged. This simple oscillator therefore provides two of the waveforms we require. Square waves can be generated if the pulse waveform triggers a multivibrator(3/3.4.2.3) and from these square waves, triangular ones may be derived using a simple integrating network as shown in (iv). Again, so that there shall be no curvature of the triangular wave, the time constant CR must be large compared with the period of the square wave so that C only partially charges.

Voltage control of frequency of this generator is easily effected by removing R from the circuit of Fig.8.7(iii) and feeding control current directly into the capacitor (e.g. from Fig.8.5) giving an arrangement as in Fig.8.7(v) where the greater the supply of control current, the more quickly C charges, hence the higher the frequency generated. In fact the frequency is proportional to the control current (which we recall is an exponential representation of the control voltage) and with some circuit refinements such a VCO is capable of operating over six or more octaves. The use of buffer amplifiers is to present a high impedance to the oscillator circuit to ensure that only an insignificant current is drawn and Fig.8.7(v) shows how sawtooth, pulse, square and triangular waveforms are generated from one basic stage.

Finally, through Fourier (App.2) we can also appreciate

the simplicity of generating a sine wave from any of the others by merely adjusting the VCF (Fig.8.3) so that it cuts off just above the fundamental. Shorn of all its harmonics, any wave reverts to the sine.

APPENDIX 1. MECHANICS

Mechanics is a branch of applied mathematics which deals with motion and the appendix is useful in achieving a better acquaintance with the sound wave as discussed in Sect.1.3.

A1.1 FORCE AND PRESSURE

Galileo (the Italian physicist and astronomer) was one of the early investigators on the subject of motion. He originated the principle of *inertia* which says quite simply that a body will continue in a state of rest or of uniform motion in a straight line unless some external agent disturbs it. Later Sir Isaac Newton went a stage further in examining the action of a disturbance and from both these men we now have the concept of *force*.[1/A9.2] It is a push or pull and is regarded as that which causes a change in the velocity of a body. Velocity change implies acceleration if the velocity is increasing or deceleration (negative acceleration) if decreasing.

Force is therefore capable of overcoming inertia which of any object is quoted in terms of its *mass*, the *quantity of matter*. Now mass and weight are often taken as being the same and in fact on Earth the difference may be negligible. However, being scientifically minded we must always calculate in terms of mass rather than weight, for compared with on Earth, on the Moon or high up in the atmosphere weights are much less but the mass is unchanged. Thus everywhere the force required to overcome inertia is the same provided that we work in terms of mass. This gives us the opportunity of defining force in relation to the acceleration it produces on a certain mass and a given force will therefore be constant irrespective of where it acts. The unit of force is defined as that which when applied to a mass of 1 kgm gives it an acceleration of 1 metre per second per second. The unit is the Newton (N), hence

$$F = m \times a$$

where   F = force in Newtons, m = mass in kgm and a = acceleration in m/s^2.

To get the newton into some perspective, it is equivalent to 0.102 kgm or about 3.6 ounces so we could get some feeling for it by thinking in terms of a weight of about 100 gms or 3½ ozs held in the hand, the weight felt is due to a force of about 1 N exerted by gravity. For readers with some knowledge of acoustics, one newton is equivalent to 10^5 dynes, a dyne being that force which imparts an acceleration of 1 cm/s^2 to a mass of 1 gramme.

One of the least complicated ways of verifying the relationship is by considering the action of gravity on a freely falling body, by free-fall is meant that no other forces are acting, strictly not even the friction of air. Early experiments showed that bodies falling freely from the same height reach the ground at the same time even though having different weights. Galileo is said to have used the Leaning Tower of Pisa for such experiments and he further discovered that bodies acquire more speed the farther they fall. Gravity in fact causes a freely falling body to accelerate at 9.81 m/s^2, hence the force due to gravity,

$$F = m \times a = 1 \text{ kgm} \times 9.81 \text{ m/s}^2 = 9.81 \text{ newtons.}$$

(ref. 1/A9.2).

Having discussed force at some length, a definition of pressure follows easily, it is simply force per area, the SI unit being one newton per square metre (1 N/m^2). Alternatively, the special name *pascal* (Pa, after Blaise Pascal, the French scientist and mathematician) may be used. The unit is equal to 10 dynes/cm^2, thus:

$$1 \text{ N/m}^2 = 1 \text{ Pa} = 10 \text{ dynes/cm}^2.$$

A1.2 ELASTICITY

Elasticity is the capability of a material to fully regain its original shape after some distorting influence has been applied and then removed. In technical terms we say that *stress* when applied to a body results in *strain*, that is, an alteration of shape or dimensions. For a perfectly elastic body the strain disappears when the stress is removed. The classical example in construction engineering is that of a metal beam (rolled steel joist) supporting a floor. With normal loads the beam *deflects* or bends. When the load is removed the beam restores to its previous position and shape because of its elasticity. Deflection and restoration can continue indefinitely provided that the *elastic limit* for the particular metal is not exceeded. When this happens the beam becomes deformed and stays that way. Building structures, cranes, bridges, etc. are all elastic and designed so that the elastic limit is not exceeded, if perchance this should happen a collapse is likely. Bricks and concrete do not in any way appear to be elastic, especially compared with metals, yet proof is there in the fact that for example, tall factory chimneys sway in the wind.

The compressions and rarefactions of a sound wave when acting on a material are therefore in fact an alternating stress resulting in an alternating strain within that material. However, generally the stress is of such infinitely small magnitude that reaching the elastic limit of the material is unthinkable, that is the wave is transmitted through the material by virtue of its elasticity but the material remains unchanged.

A1.3 THE INVERSE SQUARE LAW

This law applies when energy of any type (light, sound, etc.) is being radiated in all directions from a point source. The outgoing wave is spherical with the radius of the sphere constantly increasing. Because the energy is enclosed within the sphere, that received per unit area at some distant point becomes smaller as the distance from the source increases simply because the total emitted energy is spread over a larger area. Fig.A1.1 shows a source, S radiating E units of energy per second. We consider the energy being received over a unit of area at two distances from the source r_1 and r_2. At A, distance r_1 from S when the wave arrives:

$$\text{Area of surface of sphere} = 4\pi r_1^2 \text{ units}$$

(cm^2, m^2, etc., depending on the unit used for r_1)

$$\therefore \quad \text{Energy per second per unit area} = \frac{E}{4\pi r_1^2}$$

When the wave progresses further to B:

$$\text{Energy per second per unit area} = \frac{E}{4\pi r_2^2}$$

Then

$$\frac{\text{Energy per second per unit area at A}}{\text{Energy per second per unit area at B}}$$

$$= \frac{E}{4\pi r_1^2} \times \frac{4\pi r_2^2}{E} = \frac{r_2^2}{r_1^2}$$

which can also be written as

$$\frac{\dfrac{1}{r_1^2}}{\dfrac{1}{r_2^2}}$$

thus the energy at a distant point varies inversely as the square of its distance from that point.

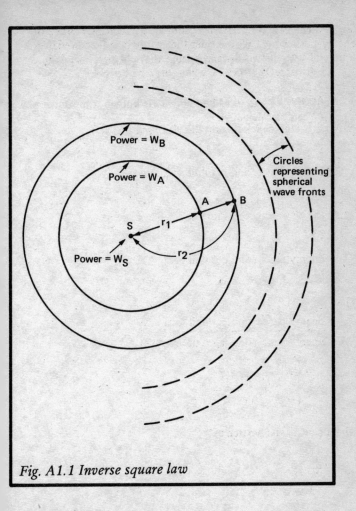

Fig. A1.1 Inverse square law

A1.3.1 Power at the Source

Energy per second we recognize as power, let that at the source be designated W_s watts and that at, for example, point A, W_A W/m² (theoretically there is no area to consider at the source).

Then

$$W_A = \frac{W_S}{4\pi r_1^2}, \quad \text{i.e. } W_S = W_A \times 4\pi r_1^2$$

and if W_A is measured and r is known, the source power W_S can be calculated.

APPENDIX 2.
ANALYSIS OF COMPLEX WAVEFORMS

The sine wave is the one of greatest simplicity because it contains no other components, it consists only of a wave at the fundamental frequency and cannot be subdivided. Expressed mathematically it is simply $v = V \sin \omega t$ where $\omega = 2\pi f$. A *complex* waveform on the other hand is non-sinusoidal although in fact it can be considered as the sum of a number of harmonically related sinusoidal waves of differing amplitudes. This is the basis of Fourier's analysis (Jean Baptiste Joseph Fourier, the French mathematician and physicist) and although it refers to continuous periodic waveforms only (i.e. all cycles are the same), it does in fact help us to make some sense of any peculiar looking waveform.

At this stage we have some experience of the addition of harmonics[(2/1.4.1)] and how a square wave is built up.[(2/1.4.2)] This Appendix provides some revision and extends the theory into other waveforms and the interpretation of Fourier equations. Surprisingly enough, although these look complicated, no further mathematical skill is required.

A2.1 EXPRESSING A SERIES

Taking the square wave as an example, it is well known to consist of a fundamental plus all the odd harmonics to infinity. Because each harmonic has its own mathematical description, the full expression for the wave must comprise an infinite number of terms. This can be handled in two ways, namely:

$$v = V \left\{ \sin \omega t + \tfrac{1}{3} \sin 3\omega t + \tfrac{1}{5} \sin 5\omega t + \ldots \right\} ,$$

the dots indicating that the series continues in the form as indicated by the terms which are written, or

$$v = V \times \sum_{n=0}^{n=\infty} \frac{1}{2n+1} \sin(2n+1)\omega t \ .$$

Σ is the Greek capital letter sigma and $\sum_{n=0}^{n=\infty}$ means the summation (i.e. the total amount by addition) of all the terms in the series from $n=0$ to $n=\infty$ where n identifies the term in the series.

The two expressions have precisely the same meaning for in the second, putting $n=0$ gives $\frac{1}{0+1} \sin(0+1)\omega t$, i.e. the first term (the fundamental) is $\sin \omega t$. Putting $n=1$ gives $\frac{1}{3} \sin 3\omega t$ which is the second term etc.

The series can also be illustrated graphically by drawing the curve of each frequency from its equation, then adding the curves together, the limitations are obvious in that the amount of work involved as more harmonics are added becomes disproportionately great. Nevertheless to picture what we do is always rewarding. A *frequency spectrum* is an alternative pictorial representation of a complex wave and what we have done so far is summed up in Fig.A2.1 where (i) shows a sine wave and (ii) and (iii) its 3rd and 5th harmonics, also sine waves according to their equations above. When we add them together as in (iv) the waveform departs very much from sinusoidal and clearly the shape of a square wave is beginning to show. As more odd harmonics are added the shape of the synthesized wave approaches the square even more and when an infinite number of odd harmonics is present the wave is truly square [this is shown up to the 13th harmonic in Book 2 (Fig.1.18), it is not repeated here but readers not possessing Book 2 can be assured that the result of adding the extra four harmonics is quite impressive]. Thus we gain the visual conviction that Fourier analysis is not just a mathematical concept but is proved in practice. As a reminder, the various scales appropriate are at the top of the Figure, for consistency we ourselves use ωt (radians).(2/1.2.4)

300

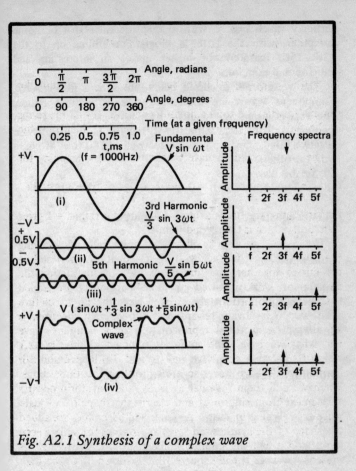

Fig. A2.1 Synthesis of a complex wave

A2.2 FOURIER EQUATIONS

Readers who would like to gain greater intimacy with the analysis may wish to plot their own waveforms from the Fourier equations given. This is quite possible using trigonometrical tables and a simple calculator, working at say, 10° intervals and at smaller ones where necessary to determine awkward changes in the curve. This may be tedious but certainly much less so with a scientific calculator or home computer, with the latter a programme adding up to the 13th–15th harmonic or higher makes an interesting and enlightening exercise.

The waveforms discussed below are shown in Fig.A2.2. Component waves are sometimes sine, sometimes cosine. The ears is insensitive to differences in phase hence we can ignore this distraction. Similarly with harmonics which have a negative value, they are just as effective but $180°$ out of phase with the others.

(i) Square Wave

Fourier equation: $\dfrac{4}{\pi} \displaystyle\sum_{n=0}^{n=\infty} \dfrac{1}{2n+1} \sin(2n+1)\omega t$

from which

1st term, $\quad n = 0$: $\quad \dfrac{4}{\pi} \sin \omega t$

2nd term, $\quad n = 1$: $\quad \dfrac{4}{\pi} \cdot \dfrac{1}{3} \sin 3\omega t$

3rd term, $\quad n = 2$: $\quad \dfrac{4}{\pi} \cdot \dfrac{1}{5} \sin 5\omega t \ldots$ etc.

and when $n = \infty$ the value is zero because

$$\dfrac{1}{2n+1} = 0 \,.$$

The square wave which has equal excursions above and below the axis can therefore be analysed into a fundamental sine wave of the same frequency with maximum amplitude $4/\pi$ (i.e. 1.27 times the amplitude of the square wave) plus all *odd* harmonics to infinity having progressively lower amplitudes as the harmonic number increases.

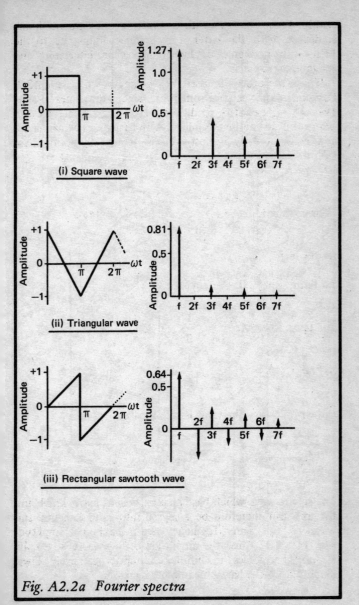

Fig. A2.2a Fourier spectra

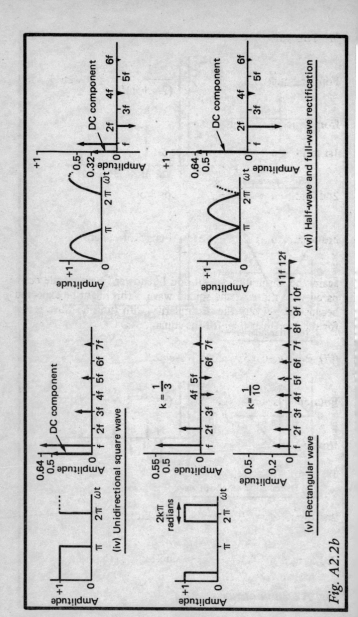

Fig. A2.2b

(ii) Triangular Wave

Fourier equation: $\dfrac{8}{\pi^2} \displaystyle\sum_{n=0}^{n=\infty} \dfrac{1}{(2n+1)^2} \cdot \cos(2n+1)\omega t$

from which

1st term, $n = 0$: $\dfrac{8}{\pi^2} \cdot \cos \omega t$

2nd term, $n = 1$: $\dfrac{8}{\pi^2} \cdot \dfrac{1}{9}\cos 3\omega t$

3rd term, $n = 2$: $\dfrac{8}{\pi^2} \cdot \dfrac{1}{25} \cos 5\omega t \ldots$ etc.

again the harmonics are all odd but lower in amplitude compared with those of the square wave. This might be expected because graphically the dissimilarity with the sine wave is less for the triangular than for the square.

(iii) Rectangular Sawtooth Wave

Fourier equation: $\dfrac{2}{\pi} \displaystyle\sum_{n=0}^{n=\infty} (-1)^{n+1} \cdot \dfrac{1}{n}\sin n\,\omega t$

from which

1st term, $n = 1$: $\dfrac{2}{\pi} \sin \omega t$

2nd term, $n = 2$: $-\dfrac{2}{\pi} \cdot \dfrac{1}{2} \sin 2\omega t$

3rd term, $n = 3$: $\dfrac{2}{\pi} \cdot \dfrac{1}{3} \sin 3\omega t \ldots$ etc.

our first example of a waveform containing both even and odd harmonics.

(iv) Unidirectional Square Wave

Fourier equation:
$$\frac{1}{2} + \frac{1}{\pi} \sum_{n=0}^{n=\infty} \frac{1-(-1)^n}{n} \sin n\omega t$$

note the d.c. component ($\frac{1}{2}$) which is added to the waveform:

1st term, $n = 1$: $\dfrac{2}{\pi} \sin \omega t$

2nd term, $n = 2$: 0

3rd term, $n = 3$: $\dfrac{2}{3\pi} \sin 3\omega t \ldots$ etc.

odd harmonics only, together with a steady component.

(v) Rectangular Wave
This is also the familiar *pulse train* where k is the *duty cycle*, that is, the proportion of the time during which the amplitude is maximum.   The pulse therefore has a width on the graph of Fig.A2.2(v) of $2\pi \times k$ radians.

Fourier equation: $k + \dfrac{2}{\pi} \displaystyle\sum_{n=0}^{n=\infty} \cdot \dfrac{1}{n} \sin nk\pi \cdot \cos n\omega t$.

There is a steady component k to which is added the wave-form:

1st term, $n = 1$: $\dfrac{2}{\pi} \cdot \sin k\pi \cdot \cos \omega t$

2nd term, $n = 2$: $\dfrac{2}{\pi} \cdot \dfrac{1}{2} \sin 2k\pi \cdot \cos 2\omega t$

3rd term, $n = 3$: $\dfrac{2}{\pi} \cdot \dfrac{1}{3} \sin 3k\pi \cdot \cos 3\omega t$

cosine waves of both odd and even harmonics but obviously not all of them because at some value of k, $\sin nk\pi$ becomes 0.

This is an especially useful waveform for electronic music generation because by altering the duty cycle, large variations in harmonic content are obtained. Suppose $k = \frac{1}{3}$, then

1st term, $n = 1$: $\dfrac{2}{\pi} \cdot \sin\dfrac{\pi}{3}\cos \omega t = 0.55 \cos \omega t$

2nd term, $n = 2$: $\dfrac{2}{\pi} \cdot \dfrac{1}{2}\sin\dfrac{2\pi}{3}\cos 2\omega t = 0.28 \cos 2\omega t$

3rd term, $n = 3$: $\dfrac{2}{\pi} \cdot \dfrac{1}{3}\sin \pi \cos 3\omega t = 0$

4th term, $n = 4$: $\dfrac{2}{\pi} \cdot \dfrac{1}{4}\sin\dfrac{4}{3}\pi \cos 4\omega t$

$$= -0.14 \cos 4\omega t .$$

Thus when $k = \frac{1}{3}$ the third harmonic is zero and by continuing the calculations we will find that the 6th, 9th, etc. are also zero. In general if $1/k = L$, then all harmonics with numbers exactly divisible by L disappear. This is demonstrated by the two frequency spectra for $k = \frac{1}{3}$ and $k = \frac{1}{10}$ in Fig.A2.2(v). Note also the lower amplitude but increase in the number of harmonics as the duty cycle decreases.

(vi) Rectification
Two other Fourier equations are worth noting here for half

and full-wave rectification,(2/1.4.4) see Fig.A2.2(vi).

half-wave

Fourier equation:

$$\frac{1}{\pi} + \frac{1}{2}\sin \omega t - \frac{2}{\pi}\sum_{n=1}^{n=\infty} \cdot \frac{1}{4n^2 - 1}\cos 2n\,\omega t$$

Expansion shows that (i) a component of the input wave is present; (ii) the d.c. component is $1/\pi$ (0.32) times the maximum value of the input wave; (iii) there are even harmonics only and (iv) the second harmonic is significant, higher ones much less so.

full-wave

Fourier equation: $\quad\dfrac{2}{\pi} - \dfrac{4}{\pi}\sum_{n=1}^{n=\infty}\dfrac{1}{4n^2 - 1}\cos 2n\,\omega t$

showing that (i) no component of the input wave is present; (ii) the d.c. component is double that for the half-wave rectifier and (iii) there are even harmonics only but double the amplitude of those for half-wave.